Saving the Starry Night

Patrizia Caraveo

Saving the Starry Night

Light Pollution and Its Effects
on Science, Culture and Nature

 Springer

Patrizia Caraveo
National Institute for Astrophysics
Milan, Italy

ISBN 978-3-030-85063-0   ISBN 978-3-030-85064-7   (eBook)
https://doi.org/10.1007/978-3-030-85064-7

Cover image credit: Astrophotographer Giorgia Hofer and partner admire Jupiter and Saturn shining above the Tre Cime di Lavaredo, an iconic landmark on the Italian Dolomites. The picture was taken at sunset at the beginning of October 2020, when the two planets were moving towards their Great Conjunction which happened in mid December (copyright Giorgia Hofer)

This Springer imprint is published by the registered company Springer Nature Switzerland AG
The registered company address is: Gewerbestrasse 11, 6330 Cham, Switzerland

To Giulia and her daddy
Great satellite hunters

Contents

1

The Sky as Cultural Heritage

I am an astronomer, and consider myself to be a very lucky person. Studying the sky is the second oldest job in the world, and certainly the most fascinating.

The astronomers of the past were marketing geniuses, because they managed to persuade kings and emperors that the future was written in the sky. They were so convincing that no ruler would think of making any decision without first hearing the opinion of the court astronomer. Now, we know that these were fantasies: the future is not written in the sky, which, on the contrary, holds memory of our cosmic past. In fact, with the exception of hydrogen, all of the elements of which we and everything around us are made were produced by the stars. We are stardust and, in my opinion, this explains why, since the dawn of time, humankind has been deeply fascinated by the sky.

It is no coincidence that the words 'cosmology' and 'cosmetics' have the same root. Cosmos means beauty. All

P. Caraveo, *Saving the Starry Night*, https://doi.org/10.1007/978-3-030-85064-7_1

we have to do is look up and the sky is at our disposal, without the need for a reservation or a ticket. The great celestial beauty is there waiting for us, ready to amaze us with its special effects, as happened in the second half of 2020 with Jupiter and Saturn approaching their spectacular **great conjunction** (Fig. 1.1).

Stephen Hawking used to say "Remember to look up at the stars and not down at your feet. Try to make sense of what you see and wonder about what makes the universe exist".

I really think we should take his advice!

In fact, there are many ways to enjoy the beauty of the starry sky. We can admire it and let our imagination run free to populate it with the myths of our culture, or, while continuing to admire it, we can study it so as to understand the laws that govern the Universe.

Since fantastic-mythological use has been with human kind since its origins, let's try to compare the celestial stories, examining the meaning of the same constellations as seen by different cultures. Depending on the place and the epoch we consider, the stars tell stories that are sometimes similar, sometimes very different, involving loves and betrayals, escapes and reunions, battles and hunts, heinous crimes and sublime actions. Orion, with the three beautiful stars in its belt, is an example of the multitude of possible interpretations. Aside from the belt of the womanizing hunter who culminates in autumn-winter, during the hunting season, the three stars represent the wrist of a hand for the Lakota Indians, a canoe with three fishermen for the Australian aborigines, the primordial turtle for the Mayas, and the firelighters for the Aztecs. Not a bad example of varied storytelling, and I left out the more gory and less edifying tales that might induce the supporters of political correctness to change the name of the constellation that, despite having its origins in Greek mythology,

Fig. 1.1 Spectacular view of Jupiter (the brightest source) and Saturn (to the left) above the Three Peaks of Lavaredo in the Italian Dolomites. Astrophotographer Giorgia Hofer and her partner are also in the picture taken at sunset at the beginning of October 2020. This image has been chosen to be the APOD (Astronomy Picture of the day) of Oct.20, 2020

is, in fact, an example of celestial globalization, since all but one of its stars have Arabic names (the exception being Bellatrix, a small concession to Latin).

Let's move on to the Pleiades, an example of "cosmic" coincidence, since the myth of the seven fleeing sisters, pursued by a hunter, is found in very distant cultures that never had occasion to come into contact before the last few centuries. In Greek culture, they are the daughters of Atlas who, pursued by Orion, were transformed into stars, but for the Iroquois, they are children who, mistreated by their parents, fled into the sky. For the Aztecs, they were at the centre of a terrible (and very complicated) fratricidal story for which the population must have had a particular predilection, because the largest temple in Mexico City was dedicated to the exterminating god. In other words, they represent a myriad of legends, generally with strong hues that often have a link to real life, as in Indonesia and Africa, where the appearance of the Pleiades was the signal to begin ploughing, while in the Andes, they were used to predict future rainfall, a method that may involve a grain of truth, since, in the presence of El Nino, the visibility of the stars is blurred.

Australians' stories are focussed on the Milky Way, which, in the Aboriginal tradition, is the path of the souls of the dead, but the clouds of dust that dominate it near the galactic centre are a symbol of life, because they are in the shape of an emu caught in the moment of hatching. This is a celestial signal: in spring, when the centre of the Galaxy becomes visible in Australia, it is time to start collecting emu eggs, so important for the survival of tribes that live in territories with very few resources.

Countless nights of patient observations with the naked eye, together with amazing intuitions, allowed our ancestors understand that celestial bodies can be divided into two classes: the stars and the planets, true celestial wanderers whose continuous movements follow clear periodicities that can be used to anticipate celestial phenomena. The ability to make predictions takes us into the field of astronomy, the oldest science and the only one to have a muse: Urania.

Apart from being fascinating, astronomy has been a useful science for millennia. The Sun, Moon and stars have served as humankind's clock and calendar, and have also pointed travellers in the right direction. Basic astronomical knowledge was fundamental to everyday life, and people were very familiar with the heavens. Considering the centrality of Sun worship in ancient megalithic civilizations, astronomer Fred Hoyle, in 1966, was the first to propose that the impressive Stonehenge circle was an elaborate solar observatory capable of predicting eclipses.

Archaeologists have learned their lesson, and now always think of astronomy when they come across large constructions. Indeed, all civilizations have built solar observatories to follow the path of the rising point the Sun that, every semester, moves between two extremes, known as the summer and winter solstices.

The most recent discovery refers to a settlement at Aguada Felix in the Tabasco region of Mexico, where, using the return signal of a laser mounted on an aircraft, the presence of a large platform, 1400 m long, 400 wide and 10 m high, was revealed under a thick blanket of vegetation. Close to it, just in the middle of the long side, there is an observatory pyramid. Archaeologists call such configurations E structures, because they are oriented towards the EAST in such a way that, from the observatory pyramid, during the solstices, you can see the Sun rising at both ends of the imposing platform, made of earth and clay; the one at Aguada Felix is certainly the largest of the many E structures mapped by archaeologists in the region. What is most striking is the age of the platform, dating back to 1000 BC. It was clearly built prior to the settlement, as if to say that the ceremonial buildings were the starting point for the Mayan cities.

Even the great festivals that marked the passing of time in ancient pagane cultures (and that, in many cases, have survived to our days) have always had an explicit astronomical reference: we say farewell to the Sun, hoping that it will come back, then we celebrate its return, grateful that it keeps the promise of a new season of abundant harvests.

Teaching of astronomy was one of the pillars of medieval culture, in which astronomy, together with mathematics, geometry and music, was part of the arts of the quadrivium, a fundamental stage in the education of Middle Age few scholars. In fact, the use of astronomical references was very common in the writings of the time. In the Italian cultural panorama, the work that best describes the close relationship between humans and the heavens is undoubtedly *The Divine Comedy*, a poetic journey that unfolds across Hell, Purgatory and Paradise. Written seven centuries ago by Dante Alighieri, *The Divine Comedy* is rich in astronomical references, which are used to indicate the time of day, the season of the year, and the direction to follow, showing that the sky was the reference system upon which everyone relied.

Not that the motion of the planets was always easy to explain. Mars, for example, occasionally changes the direction of its motion and turns back. Understanding the retrograde motions of the planets is anything but easy, especially if we assume that we, the observers of the sky, are at the centre of everything. This is called anthropocentrism, and it hides deep down in all of us. Whether we want it or not, we continue to think that we are at the centre of the system, even if this position complicates the geometric vision of the celestial spheres.

The Scientific Revolution Starts with Astronomy

In order to explain how the planets revolve around us in such a complicated way, astronomers built a wonderful mathematical model in which a planet describes a circle the centre of which moves along another circle that is centred on us. This is what we call the Ptolemaic model, and it held sway for over a thousand years until Copernicus made a portentous conceptual leap by managing to simplify the model of planetary motion. However, this simplification came at a price: the centre of the planets' motion was no longer us, but rather our star. Copernicus turned the tables and proposed the heliocentric system, which is what we studied in school. Around the Sun rotate, in order, Mercury, Venus, the Earth, Mars, Jupiter and Saturn, that is, all the planets visible to the naked eye. The Copernican system is simpler and more elegant than the Ptolemaic one, but it has some collateral problems of a political-religious nature. The Ptolemaic, earth-centric view was apparently in agreement with the biblical narrative, and any change in astronomical interpretation necessarily had religious implications. Copernicus, who was a canon in Frombork, Poland, had invested decades of his life writing De Rivolutionibus Orbium Coelestium (where the word "revolutions" refers to the orbits of the celestial bodies that revolve around the Sun), a book containing his new view of the cosmos. Apparently, Copernicus was not anxious to publish the treatise, which saw the light of day in 1453, thanks to the insistence of his disciple Rheticus, when the author was nearing his end. Legend has it that Copernicus saw a copy of the book on his deathbed. Although it expounds a vision contrary to orthodoxy, the text did not arouse the interest of the Inquisition. Perhaps the difficulty of the treatment combined with a sibylline preface (not by Copernicus) shielded him. To transform a new mathematical model describing the orbits of the planets into a cultural revolution, it would take the technological leap of Galileo Galilei, who, in 1609, built his first optical instrument.

Galileo had adapted the "device for observing at a distance" invented by the Dutch spectacle-maker Hans Lipperhey, who, sensing its potential, had tried in vain to patent his brilliant idea in 1608. The request was refused because the commission doubted that it would be possible to keep the combination of lenses at the base of the device a secret. In fact, the news began to spread throughout Europe, eventually reaching Galileo, a professor in Padua, who realized that he had a trump card in Murano glass. While, in Europe, they were combining spectacle lenses, he could use bigger (and better) lenses and obtain an instrument of superior performance. When, in August 1609, it reached eight magnifications, he convinced the Doge and the dignitaries to climb up to St. Mark's bell tower to see what the "cannocchiale" (a word he invented by combining 'cannone' ('tube')- with 'occhiale' ('glasses')) could mean for a maritime power like Venice. The Doge immediately understood the importance of the instrument and confirmed Galileo's chair for life, also granting him a significant increase in salary.

Galileo continued to improve his device until it reached 20–30 magnifications: a ridiculous number by today's standards, but, at the time, it was unrivalled. No one in Europe had anything like it.

He himself, speaking of his instrument, says that it was

..so excellent, that the things seen by means of it appear almost a thousand times larger and more than thirty times closer than if they were looked at with the natural faculty alone.

In the autumn of 1609, after meeting with the Doge, Galileo pointed his instrument at the sky and revolutionized astronomy.

He described these extraordinary observations in a booklet entitled Sidereus Nuncius, published in Venice in 1610. Despite the ambivalence of the Latin word 'nuncius', which means both message and messenger, Galileo does not pose as a celestial messenger; he is rather conveying the message of the stars that, thanks to the help of the new instrument, he was able to decode.

Unlike Copernicus' book, Sidereus Nuncius is written in a simple and very clear way, with an excellent set of illustrations. It does not describe models, but rather observations that can be easily verified by anyone equipped with a similar instrument. However, the implications are truly revolutionary. While the mountains on the Moon contradict the Aristotelian theory of the spherical perfection of celestial bodies, the discovery of the Medicean planets, orbiting Jupiter, undermines, from its very foundations, the idea that all celestial bodies must rotate around the Earth, the only centre of the Ptolemaic system.

And that was just the beginning: in rapid succession, Galileo observed the spots on the surface of the Sun and the phases of Venus that represent the proof of the correctness of the Copernican theory, because only a planet moving *within* the orbit of the Earth around the Sun can have the phases.

Supporting Copernicus, the message of the stars put an end to the geocentric theory, causing serious collateral damage to Galileo, who had to face the Inquisition. However, nothing (and nobody) could stop the Copernican revolution: the Earth was no longer at the centre of the Universe and was destined to become a smaller and smaller entity in a cosmos that, thanks to new and more powerful instruments, was getting bigger and bigger. As time went by, the solar system became more populated, with satellites around Jupiter and Saturn, periodic comets that describe long elliptical orbits, new planets. Thanks to spectroscopy, the strengths of physics, astronomy and technology joined hand in hand to provide a powerful method for understanding what stars are made of, starting with our Sun. While trying to map the geometry of our Milky Way, astronomers had to surrender to the evidence that our solar system is not at all in a central position. A century ago, bigger telescopes unveiled many other galaxies that are moving away from us. Einstein did not believe it at first, but it is precisely his general relativity that describes space-time expanding, and dragging galaxies along. In recent decades, radio astronomy, which we will touch upon in one of the subsequent chapters, and space science, born after the conquest of space, have developed.

Putting instruments into orbit has allowed us to study wavelengths that are absorbed by the layer of gas that allows us to live, but also to improve the performance of optical telescopes freed from the interference of our turbulent atmosphere. Indeed, also thanks to orbiting telescopes, it has been possible to infer the presence of thousands of other planetary systems.

This was a revolutionary discovery that took away our last glimmer of uniqueness. The Sun is one star among hundreds of billions of other stars in our galaxy, which is one of hundreds of billions of galaxies in the observable universe.

Astronomical Nobels

Astronomy is undoubtedly enjoying a golden moment. The results are so spectacular that, in both 2019 and 2020, the Nobel Prize in Physics went to astronomical research.

In 2019, the Nobel was split between the origins of the Universe, with an 80-year-old James Peebles finally seeing recognition of his contribution to our understanding of the cosmic background noise, a remnant of the Big Bang, the discovery of which had earned Arno Penzias and Robert Wilson the Nobel in 1978, and the discovery of extrasolar planets by Michel Mayor and Didier Queloz, respectively, a professor and a PhD student, who, in 1995, published their observation of a planet of Jupiter-like mass orbiting the star 51 Peg. Unlike Peebles, the two Swiss astronomers only had to wait for 24 years and the discovery of thousands of new extrasolar planets to see their extraordinary achievement recognized.

In 2020, the Nobel Prize in Physics went to black holes, with half of the prize going to Roger Penrose for his fundamental theoretical contributions in the 1960s and the other half to two observational astronomers, Reinhard Genzel and Andrea Ghez, the leaders of two competing groups who have devoted well over a decade to the study of the motion of stars in the direction of the galactic centre. Thanks to their work, which required technological development, assiduous observations and accurate data analysis, we know that the orbits of stars closest to the galactic centre trace the presence of a supermassive black hole in the centre of our galaxy.

We have this maxi black hole to thank for the fact that, at long last, the Nobel Committee seems to have begun to notice the existence of female astronomers. Andrea Ghez is the fourth woman to receive the Nobel Prize for Physics, and the first in the field of astronomy.

Curiously, both the discovery of extrasolar planets and that of the black hole at the centre of our Galaxy are based on indirect measurements. We assess the disturbance that something we cannot see causes to a celestial body. It is a method of investigation with extraordinary possibilities that, among other things, allows astronomers to map the mysterious dark matter, so pervasive and so elusive.

Despite these exciting findings, our civilization maintains a relationship with the sky that is perplexing in its dysfunction. While it is undoubtedly true that, today, astronomical knowledge has reached levels that were unimaginable only a few decades ago, the general public does not know even the rudiments of astronomy.

The Sun continues to rise and set, the days get longer and shorter, the Moon dances around the Earth, the stars move in the sky, exactly as they did thousands of years ago, but we don't seem to notice anything anymore.

And yet, the public loves astronomy: NASA counts tens of millions of links to sites with Martian photos every day. Each new mission pulverizes the links records of the previous one, but, nonetheless, a large part of humankind has lost the direct, visual relationship with the sky.

Of course, it's not all our fault. It's hard to love something we can't see. The starry vault gets paler every year because of the light pollution that we create when we try to cancel the night. To overcome our primordial fear of the dark, we turn off the stars, thus missing one of the most exciting visions that nature offers us. It is not for nothing that UNESCO has decreed that the starry sky is a part of humanity's heritage, to be defended so that we might hand it down to future generations. It is not an impossible mission: with a little attention, we can manage to light the way without interfering too greatly with the darkness of the sky. It is an effort that astronomers carry out in partnership with the public, because the sky belongs to everyone and everyone can contribute to its preservation.

Dante ends his journey in the Inferno with a vision of the stars:

E quindi uscimmo a riveder le stelle

Which, following Henry Wadsworth Longfellow's translation of Dante's Divine Comedy, becomes.

Thence we came forth to again behold the star.

A reassuring presence that represents a message of hope.

It is my sincere hope that we will always be able to look up and see the stars again.

2

The Charm of Light

Light pervades our lives. Nothing around us could exist without light. Almost every form of life that has evolved on planet Earth has extracted the energy necessary for its survival from sunlight, or has fed on organisms capable of doing so by exploiting the riches offered by the soil. Life on Earth and light are very tightly linked, simply because life would not be possible without light.

The fascination that light exerts on physicists has ancient roots. The first treatise on optics was written about thousand years ago, by Ibn al-Haytham, an Arab scientist. In addition to wondering how light beams behave, physicists (and philosophers) have often pondered the very nature of light.

© The Author(s), under exclusive license to Springer Nature
Switzerland AG 2021
P. Caraveo, *Saving the Starry Night*,
https://doi.org/10.1007/978-3-030-85064-7_2

Wave or Particle?

Revisiting the atomistic theory elaborated by the Greek philosophers who were followers of Epicurus three centuries before Christ, in 1637, René DesCartes proposed that light was composed of tiny corpuscles. His contemporary Pierre Gassendi was of the same opinion, but the corpuscular theory found its true champion in Isaac Newton in the last decades of the 1600s.

The end of the century that began with the discoveries of Galileo was an extraordinarily fertile period for studies on the nature and speed of light, a quantity that Galileo had considered measuring, with an experiment that, however, he failed to carry out. While Galileo had imagined using the transit time between two hills, Ole Roemer, in 1676, noticed some oddities in the eclipse entry times of Jupiter's satellites. Since the period of the satellites was known, when they passed behind the body of Jupiter, their time of ingress and egress was used as a clock to allow for the calculation of longitude at sea. Roemer understood that the oddities were due to the mutual motion of the Earth and Jupiter. When the Earth is closer to Jupiter, light must travel a shorter distance than when (six months later) the Earth is on the other side of the Sun and is, therefore, further away. The anomaly of the eclipses of the Medicean satellites led to the first measurement of the speed of light.

The value obtained is reasonably close to the correct one and did not arouse controversy. The same thing cannot be said for the nature of light, which was the subject of a scientific clash between two giants. While Isaac Newton was a dedicated supporter of the corpuscular theory, believing that light is composed of tiny corpuscles that travel in a straight line and can be reflected on surfaces, Christiaan

Huygens advocated for its undulatory nature, according to which light is a wave. This assertion has an immediate consequence that did not escape Huygens. Since it is well known that sound waves need a medium within which to propagate, light too had to propagate through something present everywhere, something called ether. Both scientists brought evidence in favour of their assertions, which were described in extensive treatises. Huygens, who was Dutch but worked in Paris with Giovanni Domenico Cassini, published **De la lumière** in 1690. However, Newton was certainly not to be outdone, presenting his monumental **Philosophiae Naturalis Principia Mathematica**, known as the Principia, which first appeared in 1687, and then in later editions in 1713 and 1726, in which he disputed the presence of a medium that fills the cosmos, because it would interfere with the motion of the planets. Similarly to De Revolutionibus, the Principia are rather difficult to read, so much so that malicious tongues report a venomous comment by a Cambridge University student who, seeing the master passing by, said, *There goes the man that writt a book that neither he nor anybody else understands.*

Both theories had their strong and weak points and could have been falsified thanks to the predictions they made about the speed of light in different media, but the instruments of the time were incapable of reaching the required precision.

The answer came in 1801, when Thomas Young carried out an "epoch-making" experiment that put an end to the diatribe on the nature of light. He came up with the idea while watching ducks swimming one after the other in a pond. Each created waves, and when the waves from the motion of one duck met those produced by another, the superposition effect radically changed the pattern of the waves, with effects that were either constructive or destructive. He had observed an example of interference, which he then thought of applying to light. Thus, the two-slit experiment was born. Young arranged for light to pass through two slits, creating two light sources to be projected onto a screen. If the light were corpuscular in nature, the two slits would be seen, but if it were an undulatory phenomenon, the two slits would become sources of waves that would interfere with each other, creating a sequence of light and dark areas, known as the interference fringes. Indeed, the fringes appeared strong and clear, in unequivocal support of the wave theory. This finding was not easily accepted by the scientific world; Young himself found it difficult to accept the idea that Newton was wrong. However, the result of his experiment, although very clear, did not answer all of the questions. What kind of waves were they? Did they need the ether to propagate?

The road towards the solution of the mystery was opened in the mid-1800s by Michael Faraday, who hypothesized that light was an electromagnetic vibration of high frequency that could propagate even in the absence of a medium. It was only the first step in a process that was consolidated by James Clerk Maxwell, who, starting from heterogeneous observations, succeeded in unifying electricity and magnetism, elaborating a system of equations on the basis of which electricity and magnetism could give origin to electromagnetic waves. Moreover, He clarified that, contrary to common thought, which saw light as distinct from electricity and magnetism, light is, to all intents and purposes, an electromagnetic wave. A wave that propagates without the need for a medium, as demonstrated in 1881 by Michelson with a result confirmed in 1887 in collaboration with Morley. After two centuries, the ether hypothesized by Huygens disappeared from the scene.

The 1900s started with the revolutionary revival of the corpuscular theory by Max Planck, who suggested that, although light is a wave, energy exchanges do not take place in a continuous form: waves can exchange energy only in packets (the value of which depends on the frequency of the wave). Thus, the quantum hypothesis was born, and the following year, it was applied to atoms, which can absorb or emit energy only in discrete quantities: quanta.

This was an idea with profound implications that began to be appreciated when, in 1905, it allowed Einstein to explain the photoelectric effect.

In the end, light is both a wave and a particle (light quanta are called photons) and, depending on the case, can be described with a mathematical approach that fits either nature. This extraordinary insight earned Planck the Nobel Prize in Physics in 1918.

Light and Gravity

Light also has a close relationship with gravity, seen as the manifestation of the curvature of space–time according to the theory of general relativity published on November 25, 1915, in Berlin, where Einstein was director of the Institute of Physics at von Humboldt University.

With Europe divided by World War I, scientific papers published in Germany struggled to cross borders, but news of Einstein's new work could not go unnoticed. The physicist Willem De Sitter, who worked in the Netherlands, a neutral country, sent a copy of the paper to the Royal Astronomical Society and, in 1916, Arthur Stanley Eddington, very brilliant professor at Cambridge University and director of the observatory, had the opportunity to read it and was thunderstruck. It was a difficult theory, but it explained the motion of Mercury's perihelion, which had always eluded Newtonian theory.

Eddington immediately began to teach the new theory and to spread it within the Anglo-Saxon scientific press, becoming the English voice of relativity. That same year, 1916, he had already described general relativity to a meeting of the British Science Association, underlining the necessity to verify the correctness of the theory through the measurement of light's gravitational deviation. The presence of a mass influences the propagation of light by deviating the photons' trajectory. It is an effect already foreseen by the Newtonian theory, but general relativity amplifies it, making this parameter a key factor in falsifying or confirming the new theory. It was thought to that the Sun could be used as a deviating mass, exploiting the few minutes of darkness in a total eclipse when the Moon "turns off" the Sun and the stars appear in the sky. By comparing the undisturbed positions of the stars in a normal night sky with those measured during the eclipse, the gravitational deviation could be assessed. Despite the raging war, the Royal Astronomical Society began planning two expeditions to exploit the eclipse of May 29, 1919, which would be visible from Prince Island, off the coast of Africa, and from Brazil. It was a very suitable eclipse for such an experiment. Firstly, the Sun would be close to the Hyades, a group of bright stars, and secondly, the eclipse would be very long, with 6 min of darkness during which time astronomers would be able to expose several photographic plates so as to capture the position of the stars and measure whether they had moved, and by how much. The idea of organizing two (or more) expeditions, instead of just one, was a common astronomical practice used to avoid nasty surprises due to the weather's vagaries when one plans to observe a unique event that would not be repeated. Indeed, having multiple observation sites far apart increased the chances that at least one would enjoy good weather that would allow the desired measurement to be obtained. Fortunately, the war ended and the expeditions became a reality. Eddington left in March so as to be ready by the end of May. The eclipse was expected to take place in early afternoon, and the morning at Prince Island was rainy. The clouds opened just in time, but the sky remained partly cloudy. Despite the less than perfect weather, Eddington tried to get as many photographic plates as possible, hoping that, in at least one of them, the stars would not be covered by clouds. Developing the plates in the following days, he realized that his wish had been granted. On June 3, he wrote in his notebook "… *one plate I measured gave a result agreeing with Einstein*". To those who, years later, congratulated him for having been able, despite the dark years of the war, to plan such a daring expedition, Eddington replied that he deserved little credit since he had been so sure of the accuracy of the theory that he had never really felt the need to verify it. The results were presented at a legendary meeting of the Royal Society on November 6, 1919, and the news made the front pages of all the newspapers, giving Einstein worldwide fame.

Nowadays, the use of gravitational deviations of light by celestial bodies that, by chance, happen to be aligned with its propagation path is a very important tool for understanding the distribution of matter in our Universe. When the optical emission of a galaxy is deflected by another galaxy, or by a cluster of galaxies, there is an amplification of the original flux, thanks to an effect known as gravitational lensing. Depending on the reciprocal positions of the source and the deflecting mass, multiple images can be formed, sometimes conspicuously deformed by the lensing effect (Fig. 2.1).

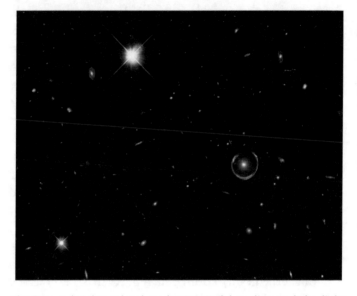

Fig. 2.1 A bright red galaxy (LRG 3-757) has distorted the light of a much more distant blue galaxy. Thanks to the fortuitous—but perfect—alignment between the positions of the two galaxies, the light of the more distant one has been distorted to form a circle called an Einstein Ring that stands out in this image obtained by the Hubble Space Telescope (*Credit* ESA/Hubble & NASA)

By mapping the distortions, we can reconstruct the mass distribution that is responsible for the effect without distinguishing between visible and dark matter. In this way, the light also allows us to see the dark matter that has mass but does not emit anything (Fig. 2.2).

And that's not all… Gravitational amplification allows us to reveal ancient galaxies that, without this assistance, would be too weak to be "seen" by our instruments. This is the case with Tayna, a galaxy born just 400 million years after the Big Bang and revealed by the Hubble Space Telescope thanks to the amplification provided by a cluster

Fig. 2.2 A cluster of galaxies about 4.6 billion light-years away from us acts as a gravitational lens for the galaxy dubbed Sunburst Arc, 11 billion light-years away. The four arcs seen in the Hubble Space Telescope image are multiple replicas of the same galaxy appearing at least a dozen times, turning Sunburst Arc into the brightest galaxy seen through lensing (*Credit* ESA/ Hubble & NASA)

of galaxies in the middle. Gravitational lenses can also be time machines, because, by deflecting the trajectories of light, they can lengthen the path that photons must travel to reach our instruments. In this way, it has been possible to observe the same unique phenomenon, such as the explosion of a supernova, as if it were occurring several times. A touch of magic from the combination of light and gravity (Fig. 2.3).

Supernova Refsdal ▪ Galaxy Cluster MACS J1149.6+2223
Hubble Space Telescope ▪ ACS/WFC ▪ WFC3/IR

NASA and ESA   STScI–PRC15–08a

Fig. 2.3 Another image from the Hubble Space Telescope showing a bright (and massive) galaxy at the center of the MACS
cluster J1149.6+2223 distorting a more distant galaxy in which
we see a supernova explosion taking place four times (*Credit*
NASA-ESA)

Visible and Invisible Light

Light is beauty, it is science, it is energy, but it can be so
much more.

Stonehenge and the great prehistoric monuments tell us that our ancestors assigned divine properties to light. This tradition has not been completely lost: astronomers are the modern worshippers of light, since almost all astronomical research is based on the study of some form of light.

Optical astronomy has allowed us to make giant steps towards understanding planetary systems, the life and death of stars, the formation of galaxies and how they move away from each other, following space-time expansion. By studying the speed of stars and galaxies, we can also measure mass that is there but cannot be seen, the dark matter. This is a cumbersome presence, far greater than that of "normal" matter we are made of, but of which we still know too little.

The "visible" radiation, however, carries only a small part of the information produced by celestial objects.

Indeed, it accounts for a tiny fraction of the electromagnetic spectrum that extends towards longer and shorter wavelengths, invisible to our eyes (Fig. 2.4).

The conquest of the invisible is the result of scientific and technological progress, because, in order to measure the emission of a celestial object beyond the visible, instruments suitable for detecting the different wavelengths are needed. However, having detectors designed to measure every type of radiation is not enough to give us access to all of the information contained in the electromagnetic spectrum.

The emission from celestial bodies has to contend with the Earth's atmosphere, a shell of a few tens of kilometres of gas that acts as a selective filter, allowing some types of radiation to pass through and stopping others. In fact, looking at the graph of the transparency of the atmosphere, we realise that a large percentage of the radiation does not reach the ground because it is absorbed. The only

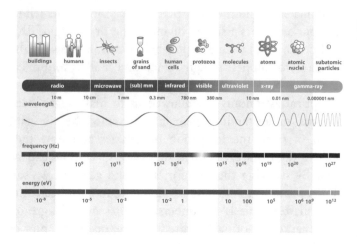

Fig. 2.4 The electromagnetic spectrum. To describe any type of radiation, we can use the wavelength (above), the frequency (i.e., the number of cycles per second) or the energy of the corresponding photon (the scales below). In the radio range, we speak of the frequency of the radiation, in optical and infrared, we prefer to use the wavelength, while in X and gamma astronomy, we always refer to the energy of the photons (*Credit* ESA/AOES Medialab)

frequency intervals that can overcome the barrier of the atmosphere are the radio and optical ones (Fig. 2.5).

We should certainly not be surprised to discover that our atmosphere is transparent to visible radiation, the one to which our eyes are sensitive. After all, we evolved on planet Earth and have adapted to its environmental conditions. However, we should not forget that the "visible" part is really a very small fraction of the electromagnetic spectrum, just 0.0035%. Much more extensive (in frequencies and/or wavelengths) is the transparency to radio waves, a necessary prerequisite for the success of radio astronomy, derived from radar studies developed during World War II.

Optical and radio astronomies are the only ones that can be executed through the use of instruments firmly

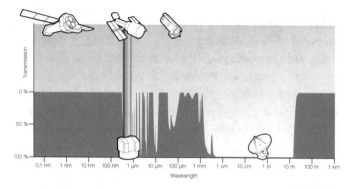

Fig. 2.5 Atmospheric transmission ranging from 0% (atmosphere is completely opaque) to 100% (atmosphere is transparent). (ESA/Hubble F. Granato)

anchored to the ground, and that is why they will be the focus of our attention.

This is not to say that those areas of astronomy that rely on satellites above the atmosphere to catch infrared, ultraviolet, X-ray and gamma rays are less important. Far from it.

The last half century has seen great successes in this field, and it is now standard practice to study celestial objects at different wavelengths so as to understand the physical mechanisms responsible for their behaviour.

In most cases, the process responsible for the emission that we record is thermal in nature, i.e., related to the temperature of the celestial objects. Thermal processes are what make stars shine.

But there is also another Universe, where, in environments with extreme characteristics, cosmic accelerators can produce high energy radiation that has nothing to do with the temperature of celestial objects. It is a violent Universe, where phenomena of unprecedented power take place, capable of generating photons of energy equal to millions, billions and even trillions of times those of visible light. We are talking about gamma photons, the most energetic radiation produced by celestial objects, that allows us to guess what is happening near neutron stars, stellar black holes, supermassive black holes at the centre of distant galaxies and supernovae explosions. Precisely because they are so energetic, gamma photons cannot be produced by the same thermal processes that make stars shine. Accelerated particles, either protons or electrons, are needed to produce gamma-rays through their interactions with the surrounding matter, or with magnetic fields, or with other photons. However, even the most extreme celestial bodies are not limited to producing only extraordinarily energetic radiation. An electron spiralling in a magnetic field can produce radio waves, as well as optical or X-ray emission. Therefore, in order to fully understand the behaviour of a celestial object, astrophysicists have learned to use all channels of the electromagnetic spectrum, thanks to a comprehensive approach called multi-wavelength astronomy.

X-rays are light as the glow of Venus or the radio ticking emitted by neutron stars: it is only their production that requires more extreme, violent, hotter, celestial objects. Light is the background radiation emitted by the newborn Universe and light is the laser beam that goes to the Moon and back every night to measure its distance with great precision, using the mirrors left behind by astronauts half a century ago. It is a beautiful experiment, proposed in the '60s in hope of finding some flaw in the theory of general relativity, but that ended up confirming it down to the smallest details.

The Latest Astronomical Achievements

As we have already mentioned, modern astronomy was born in 1609, when Galileo pointed his telescope at the sky for the first time and discovered that it was full of myriads of stars that could not be seen with the naked eye. It was the beginning of a great revolution: by improving its vision of the cosmos, humankind discovered that it was not at all at the centre of the Universe. For centuries, people had been lulled into the belief that all the bodies in the Solar System revolved around the Earth. Complicated mathematical theories had to be constructed to make the accounts add up, but nothing could withstand the disruptive new astronomical observations. It is the Earth that revolves around the Sun, along with all the other planets. Moreover, over the centuries, we have come to realise that our Sun is nothing special at all, just one of hundreds of billions of stars that make up our Milky Way. Fortunately, it is by no means at the centre of our galaxy, but rather occupies a remote, semi-peripheral position where no overly violent celestial phenomena occur. Just a century ago, we realised that the Milky Way is only one of the hundreds of billions of galaxies that make up the portion of the Universe that we can study. Finally, a quarter of a century ago, our anthropocentric view was definitively defeated, because we had to surrender to the evidence that our solar system is not a unique example. By measuring the "disturbance" a planet causes to its star as it describes its orbit, astronomers realised that other planetary systems existed. It was not an easy observation: the "disturbance" is tiny, but its repetitive nature (linked to the planet's orbital period) makes it recognisable. The search for extrasolar planets is one of the most fascinating chapters in contemporary astronomical research. In little more than a quarter of a century, the number of extrasolar planets went from 0 to more than 4,000, and the pace of discoveries continues to grow. We started by detecting massive Jupiter-like planets, whose presence is easier to infer near their star, but the techniques were subsequently refined, and we are now able to find planets that are not very different from our Earth.

We now know that having planets is a common feature of stars, indeed, we can say that every star in the Milky Way hosts at least one planet, some of which might not be very different from Earth. In addition, some of Earth's potential twins are in the habitable zone of their stars, i.e., they are at the right distance to receive enough energy to allow the water that may be present on their surface to be liquid (provided there is an atmosphere). Terrestrial planets in the habitable zones of their stars could even offer favourable conditions for the development of life… who knows?

But news from astronomy never ceases to make headlines: as in the case of the first interstellar visitor.

On **October 19, 2017**, a fast-moving object was discovered in the data from the Pan-STARRS wide-field telescopes. Initially classified as a comet, its hyperbolic orbit made it famous, as it was recognized as the first visitor from another planetary system. Named Oumuamua, *'messenger from afar',* in the indigenous language of Hawaii, where the Pan-STARRS telescopes are based, it began to be considered an escaped comet from another planetary system. This hypothesis was validated by measurements of its surface reflectivity, similar to that of comets, which are dirty snowballs composed of ice mixed with dust and rock and often covered with dust or very dark organic (tar-like) molecules. The interpretation faltered when no gas ejection activity and no tail formation were observed. The elongated cigar shape, vaguely similar to the spaceships from science fiction stories, together with its interstellar origin, prompted Star Trek-type scenarios. Although skeptical, astronomers pointed their radio telescopes at Oumuamua to see if it had radio emission, without finding any signal. That cooled enthusiasm. Then, Oumuamua, continuing on its trajectory, drifted away and, as of January 2, 2018, had become too faint to allow for further observations. However, its speed had been measured very carefully with the hope that the reconstruction of its trajectory would unveil useful clues as to its origin.

It has since been discovered that its path cannot be described solely on the basis of the gravitational forces coming from the Sun, the eight planets, the Moon and the larger bodies in the asteroid belt. There is a "non-gravitational" deviation that can be explained by a radial acceleration inversely proportional to the distance from the Sun, or to the square of the distance from the Sun. The authors of the measurement say that some kind of outgassing or, alternatively, radiation pressure from the Sun would be perfect to explain a radial acceleration inversely proportional to the square of the distance. Such an effect has been seen in cases involving small asteroids, however, for Oumuamua, the acceleration would imply that the object was made of very low density material, 1000 times less dense than similar sized asteroids in the solar system. And here comes the alien hypothesis again. It could be a solar sail that had escaped the control of its makers. As a first approximation, it could be an abandoned piece of space debris that, in its interstellar journey, must have become covered with dust (interstellar, specifically) that would account for the colours that have been measured. After all, we too are thinking of using solar sails to gain additional thrust in interplanetary travels by exploiting the radiation pressure of sunlight. In case the idea of an abandoned solar sail moving aimlessly through the galaxy doesn't satisfy us, we can also imagine that it was intentionally directed towards our solar system. Of course, there is no way to test these fanciful hypotheses, but why limit ourselves?

We have learned all of this by studying the sky, deciphering the messages that the stars send us through the light that we collect using increasingly larger telescopes in order to study, in ever greater detail, ever fainter and more distant objects. In astronomy, going far away means going back in time to investigate how stars and galaxies appeared billions of years ago. We wish that we could study the first galaxies that formed in our Universe. But big telescopes and futuristic instruments are not enough: we need the sky to be dark.

3

The Need for Darkness

One of the most iconic images obtained during the Apollo program is certainly the first colour photo of our planet as seen by the Apollo 8 astronauts, the first persons to reach the Moon and orbit around it. As they were emerging from the Moon's shadow, the astronauts, William Anders, Frank Borman, and Jim Lovell, saw the Earth rising in front of them. Recordings of their conversations testify their amazement at the sight. They were taking pictures of the lunar surface with a camera equipped with black and white film, but, understandably, felt the need to switch to colour film to capture the historic moment. They were the first human beings to see the sunlit Earth shining against the deep darkness of space (Fig. 3.1).

Available in seemingly unlimited quantities in the space that surrounds us, darkness has become a rare commodity for our luminous civilization, which has come to face a devious, but certainly not invisible, enemy: light pollution. It is a by-product of modern technology, combined

© The Author(s), under exclusive license to Springer Nature Switzerland AG 2021
P. Caraveo, *Saving the Starry Night*,
https://doi.org/10.1007/978-3-030-85064-7_3

Fig. 3.1 Earthrise. This is the iconic image taken by Bill Anders, one of the Apollo VIII astronauts, on December 24, 1968 (*Credit* NASA)

with our ancestral fear of the dark that pushes us to light up the night, forgetting that, to be useful, light must be directed downwards, where we live, and not upwards, where the stars shine. It is, at the same time, a source of economic waste and aesthetic damage, because, if we light-up the night, we turn off the stars, as well as doing harm to ourselves and to the planet.

It has not always been like this: at one time, the great astronomical observatories were built in the city centres, sometimes using the roofs of buildings, or in their immediate vicinity. Think of the historic Observatoire de Paris, a visionary work of the Sun King, the Capodimonte

Observatory in Naples, very close to the Bourbon Royal Palace, or the one located in Palermo on the roof of the magnificent Palazzo dei Normanni. It was while working at the Palermo observatory that, on January 1, 1801, Giuseppe Piazzi discovered Ceres, the largest asteroid in the solar system. Originally called Ceres Ferdinandea, in honour of King Ferdinand of Bourbon, for almost half a century, it was considered the eighth planet of our solar system (at that time, Neptune had not yet been discovered, while William Herschel, with his telescope, had already added Uranus to the family of planets). In Milan, too, the astronomical observatory was in the centre of the city, on the roofs of the Palazzo Brera. It was there that Giovanni Schiaparelli, at the end of the nineteenth century, made observations with his futuristic (for the time) telescope, which dominated the rooftops of a city that was still only dimly lit (Fig. 3.2).

Having gained renown thanks to his discovery of the nature of shooting stars, which he proved to be linked to the orbits of particular comets, Schiaparelli ventured, somewhat by chance, to study the surface of Mars. The details he saw on the red planet prompted him to repeat

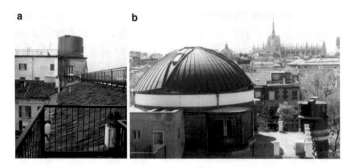

Fig. 3.2 Left Dome of the telescope used by Schiaparelli. Right: the second, more recent, dome of Brera against the spires of Milan Cathedral (*Credit* INAF)

the observations at each Mars opposition, an orbital alignment that happens every 26 months when Mars is closest to Earth. Now, we take advantage of the opposition to launch our probes to Mars, but in the late 1800s, Schiaparelli could only observe the planet and draw what he saw. The few gas lights didn't bother him and, opposition after opposition, Schiaparelli mapped Mars' surface in detail, filling dozens of notebooks with sketches of what he saw (or thought he saw) night after night. In Schiaparelli's vision, Mars was dotted with vast dark areas that he called seas, connected by more or less straight lines that he called canals. The map, published in the annals of the Accademia dei Lincei, produced a profound echo.

Schiaparelli was a famous astronomer, and his articles were widely translated. Indeed, the Martians were born owing to a translation error. Reading the word "canali" in the Italian text, Percival Lowell, an aristocrat from a rich Bostonian family with many interests, translated it into the closest word English had to offer: "canals". He would have done better to use "channel", which indicates an arm of the sea (the Channel, for example). "Canal", instead, is used exclusively to indicate an artificial structure. A mere error of translation, but one laden with consequence: if Mars is crossed by artificial channels, someone must have dug them. So, the need for an advanced civilization was born; the Martians, thus, fully deserve the label *Made in Italy*, since they are a by-product of the skies of Milan.

In the 1920s, with the dawn of electric lighting, everything changed, and the observatory at Brera had to move its observational activity to somewhere outside of the city, to Merate, where the astronomical work continued until it was again forced to surrender to the light pollution from the halo of the growing cities.

Succumbing to light pollution is a common fate for all historical Observatories built within or in close proximity to cities. The 100-inch Mount Wilson Observatory, where, a century ago, Edwin Hubble figured out that galaxies move away from each other, has been blinded by the lights of Los Angeles.

Too Many Lights Turn off the Stars

Light pollution hampers astronomy both at professional and amateur levels.

However, ordinary people suffer the most damage, because they are bound to lose all familiarity with the sky in just a short time.

It is not by chance that, in England, the sale of popular books on astronomy surged during the Second World War, when compulsory blackouts led to rediscovery of the sky's beauty. But it is not necessary to go back in time. Contemporary blackouts are just as useful for seeing the sky filled with stars.

A friend who was in Cortina, a well known sky resort on the Dolomites, during a prolungate Christmas blackout a few years ago, after telling me about his various misadventures, added: "But I saw the most extraordinary starry sky in history". The lack of lighting, obviously combined with good weather conditions, had allowed him to enjoy the beauty of the sky, free, for once, from light pollution.

Emblematic is the case of New York, where the Moon no longer appears with the same clarity that it used to. This is a situation that is, unfortunately, far too prevalent: it remains true that, from many (almost all) urban areas, it is no longer possible to enjoy the view of the starry sky. And things don't much improve when moving away from the inhabited centers.

Dust particles and aerosols in the air are very effective in scattering photons that have escaped from poorly designed lighting systems used for monuments, gardens, streets and shopping malls. This produces extensive light halos, which increase the sky's background level, making all the faintest signals disappear. In other words, darkness is never total.

This is something that you will easily notice if, in August, will go hunting for shooting stars. These are called Perseids, because the luminous trails point towards the constellation of Perseus. We know that they are grains of dust left by the comet Swift Tuttle, which, around August 10, hit our atmosphere because the Earth crosses the orbit of the comet (which passed in 1992 and will return in 2126), where the residual dust of the tail is trapped. As we have already mentioned, it is a nice result that arose from the intuition of Giovanni Schiaparelli, who rightly gained world-wide fame.

Find a dark place and make yourself comfortable, because the falling stars, even if numerous, do not arrive on command, and thus it takes patience. If the Moon is up, you will immediately notice that its light will make your search more difficult, because it increases the sky's brightness, thus decreasing the signal-to-noise ratio between the light trail you are looking for and the sky's background. Fortunately, sooner or later, the Moon will set. What will continue to keep you company, however, is the diffuse glow caused by the lights of cities, large and small, that dot our world. You will realize that light pollution is insidiously present even in secluded places, because the light from urban centres is diffused by particles in the atmosphere and forms vast halos. Look at this beautiful photo taken from the Aosta Valley, from about 2000 m above sea level (Fig. 3.3). The location is propitious, the sky is beautiful, featuring a majestic Milky Way, but you can certainly notice a suspicious glow on the left. It is the city of Turin, which is more than 100 km away, but still makes its presence brightly felt. However, halo of light aside, the Milky Way looks very beautiful, so beautiful that many of you will be wondering how it is possible that you do not remember ever seeing it like that. Manifestly, light pollution is hampering your vision of the starry night. And you're not alone—some two billion people live in regions that are too brightly lit to see the Milky Way.

It should be apparent that ours is a "luminous" civilization, but certainly not a "well-lit" one, since the criteria adopted for the lighting of urban areas seem to be dictated by humankind's perverse (and perhaps ancestral) tendency to cancel the night, rather than by the optimization of the cost/benefit ratio.

The result is an illumination, sometimes mediocre, accompanied by a waste of energy estimated at around 30%, lost in an undue, as well as useless, illumination of

Fig. 3.3 High-altitude view of the Milky Way. The photo was taken by Alessandro Cipolat Barres near Saint Barthélemy (Valle D'Aosta). Torino, 100 km away, is responsible for the diffuse brightness on the left (*Credit* Alessandro Cipolat Barres)

the sky that has the sole effect of compromising the possibilities of carrying out even the most elementary astronomical observations by most of the civilized world.

4

The Earth at Night

Each of us, looking up at night, has a subjective perception of the sky's darkness, but, to get a global vision of the
Earth at night, we need to use images collected by satellites. Under natural conditions, the non-sunlit hemisphere
should be perfectly dark, with the exception of lightnings,
natural forest fires and northern lights. Instead, the Earth
is bright!

It would be a vision of great beauty, if it were not for
the fact that it is due to the enormous waste of energy that
we are perpetrating. Let's start with a "historical" map, the
first of its kind, built with a collage of 40 images extracted
from the DMSP (Defence Meteorological Satellite
Program) archives. These are satellites in sun-synchro-
nous polar orbit (at an altitude of 825 km with a period
of 102 min) tailored in such a way that the longitude of
the satellites' equator crossing increases by 25° each orbit.
Such a particular choice of orbital parameters means that
the satellites always fly over different regions of the Earth

© The Author(s), under exclusive license to Springer Nature
Switzerland AG 2021
P. Caraveo, *Saving the Starry Night*,
https://doi.org/10.1007/978-3-030-85064-7_4

at the same local time. We are interested in the data collected by the satellites that, over the years, have covered the noon-midnight orbit. Every 0.4 s, the satellites take a picture of a region of about 3000 km in the East-West direction about 3 km thick with a resolution varying between 3 and 12 km (depending on the position within the strip). The satellite repeats the operation in two filters, visible and infrared, and provides global coverage of the Earth, always captured at the same local time, several times a day. The maps, owned by Air Force Global Weather Central, are used for weather forecasting, and are archived in Boulder after 60 days. The usefulness of the archive is not confined to meteorological studies: the data can be used for multiple purposes, from monitoring the state of glaciers to studying light pollution in the night sky.

In fact, the DMSP satellites have been the only meteorological satellites to operate at night in the blue to near-infrared range, which is particularly suitable for detecting city lighting and fires of various natures. However, this was not the original purpose of the OLS (Operational Linescan System). The satellites were supposed to pick up the Moon's reflection in the clouds to help military aircrafts navigate at night. It was subsequently realized that, on Moonless nights, the satellites were sensitive enough to see city lights. Thus, *Earthatnight* was born, and the very first image of the Earth at night was released. The photos chosen to build the collage were taken locally at midnight under cloudless and Moonless conditions in the years between 1974 and 1984. The result is fascinating from many points of view (Fig. 4.1).

In the night map of our Earth, almost all human settlements with a few hundred inhabitants (and good lighting standards) can be recognized.

Fig. 4.1 Earth at Night. Collage of 40 images taken by DMSP satellites from 800 km height locally at midnight in various locations in the years between 1974 and 1984. The only natural source of light is the aurora borealis; everything else is human-made. We see all the urban areas of the civilized world, the major thoroughfares, the fires burning in the African savannas, the controlled fires of oil fields, and the Japanese fleet, which uses lights for shrimp fishing (*Credit* D.Crawford and W.T.Robinson- ASP)

It is worth pausing for a moment to look at the collage, which represents a splendid social map of our Earth that has been used to estimate the number of inhabitants in urban areas, but also to measure the spatial extent of built-up areas in order to get an idea of land consumption. Light yields a census of population, but also a measure of wealth.

The contrast between North America, Europe and Japan and the rest of the world speaks for itself: 1/4 of the world's population consumes 3/4 of the energy, and it shows! Obviously, what we see must be normalized to the number of inhabitants of the nation. Indeed, based on this map, it was possible to compute that, in the time lapse considered, a citizen of the US consumed twice as much energy as one of Japan, 30 times more than one of China and 75 times more than one of India. Of course, the social considerations don't end there; we see the fires burning the tropical forests and those of the oil installations in the Persian Gulf. In the Sea of Japan, one cannot fail to notice the extremely powerful floodlights used by Japanese fishing fleets to illuminate the sea and bring shrimp to the surface, while the contrast between brightly lit South Korea and completely dark North Korea is stunning. In Europe, one notices the daylight illumination of Holland and Belgium against the (relative) darkness of France. In North Africa, the Nile valley shines, while in Siberia, one can follow the route of the Trans-Siberian railway. In South America, the megalopolises of Sao Paolo and Rio de Janeiro (linked by the infamous highway) stand out, along with Buenos Aires, contrasting with the legendary Manaos in the middle of the dark Brazilian Amazon forest.

After this first map, the studies continued, with new generations by DMSP satellites equipped with more sensitive instrumentation, yielding sharper, higher resolution images. These improved capabilities can be immediately appreciated looking at the image built from the 1994–1995 data. To help in identifying relevant features, a colour version was also made, with forest fires highlighted in red, gas fields in green and fishing fleets in blue (Fig. 4.2).

In 2011, the Suomi NPP (National Polar-orbiting Partnership) satellite launched by NASA in collaboration with NOAA (National Oceanic and Atmospheric Administration) and the U.S. Department of Defence was added. NASA's Black Marble project allows you to see up-to-date views of the earth at night through https:// earth.google.com/web.

Fig. 4.2 View of the Earth at night in color, with forest fires (red), gas fields (green) and fishing fleets (blue) highlighted (*Credit* NASA)

Artificial Illumination Grows Continuously

In addition to studying the current situation, with more than twenty years of satellite observations available, it is possible to study how light pollution has changed (in general, for the worse) over the years.

In fact, when we correlate the data on light pollution with other indicators, such as GNP, electricity consumption, and number of inhabitants, we find that pollution is well correlated with GNP and energy consumption. The correlation is less evident with the population census, because lighting standards depend a lot on the wealth, as well as on the customs, of the country. Indeed, if we study light pollution at the regional and local levels, there are strong discrepancies from country to country. Italy, for example, emits about three times more artificial light per capita than Germany, which has a higher individual income.

Fig. 4.3 The Earth of Night in the 2016 high-resolution version (extracted from https://earthobservatory.nasa.gov/) (*Credit* NASA)

However, where wealth is combined with a high number of inhabitants per square kilometre, the situation quickly becomes difficult (Fig. 4.3). One third of the planet's population lives in regions so flooded with artificial light that the Milky Way cannot be seen. We are talking about more than 2 billion people, most of whom reside in Europe and the United States, but we must not forget the cases of Singapore, Kuwait, Qatar, the United Arab Emirates, South Korea, Japan and Israel. If we examine the current map of the brightness of the night sky over Europe, we see two vast patches of high pollution: one corresponds to the Belgium-Holland-Northern Germany triangle, while the other covers the entire Po Valley in northern Italy. From these regions, it is impossible to see the Milky Way.

In fact, Italy enjoys the unenviable record of being one of the industrialized countries with the highest amount of light pollution. Among the G20 countries, only South Korea has worse conditions.

Illumination Rhymes with Population

Correlating satellite data, which measure upward scattered light, subdivided at the local level of provinces (in Europe) and counties (in the USA) with the data related to population density, or to wealth, it is possible to estimate the average light pollution produced by each person in the regions under examination. Comparing the situations in Europe and the USA, we notice that there are deep differences in the parameters' variation span. In Europe, for example, the most light-polluted region is Delft en Westland, in the Netherlands, which emits almost 7,000 times more light than the least polluted region in Eilean Siar in the U.K. Western Isles, while the individual contribution to sky illumination varies by a factor of 120. The reason behind the Dutch town's record is easily traceable to its huge greenhouses, where night lighting is a method of accelerating plant growth (Fig. 4.4).

Although these values are already quite impressive, the differences are even greater in the United States, where the District of Columbia emits 200,000 times more light than Yukutat City, Alaska, and the individual contribution varies by a factor of 16,000 (Fig. 4.5).

The difference in the parameters' range of the European and American maps is due to the fact that, in the United States, along with large cities, there are still many sparsely populated areas with low light pollution that one simply does not find in Europe.

It must be noted that the ranking in light emission at the regional level, although easy to obtain, is not immediately reflected in the ranking of individual contribution, because the many inhabitants of a big city, very illuminated, do not automatically go to the top of the ranking,

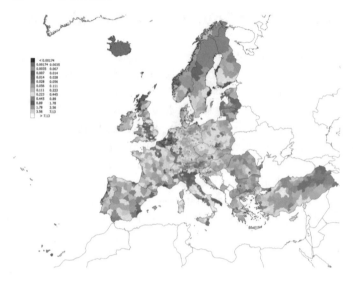

Fig. 4.4 Maps with the light directed (and dispersed) towards the zenith in Europe, organized at the local or provincial level. Courtesy of Fabio Falchi (from F. Falchi at al, Journal of Environmental Management, 248, 109227, 2019)

since, on average, the emission of each person does not turn out to be very high. For this reason, the contribution pro-capita, although interesting, especially if considered together with the data on average wealth, is a parameter that needs to be weighed carefully.

The Astronauts' Vision

Astronauts working aboard the International Space Station (ISS) love to spend what little free time they have in the *cupola* enjoying the incomparable view of the Earth below them. I'm not unveiling a secret when I say that many have been fascinated by the Earth at night, when the lights

Fig. 4.5 Maps with the light directed (and dispersed) towards the zenith in the United States, organized by counties. Courtesy of Fabio Falchi (from F. Falchi at al, Journal of Environmental Management, 248, 109227, 2019)

of the cities become the only point of reference. In addition to highlighting geographical conformation, the lights trace the urban plan of the metropolis and the technological solutions adopted for lighting. Cities at night are therefore a concentration of geography, culture, history and technology that distinguish them from one another and make them easily recognizable. The astronauts on the Space Station, who pass from blinding sunlight to complete darkness every 45 min, are very good at recognising them, using lights as a distinguishing feature. The European cities have a radial structure, while those of the American West are squared with streets in a north-south and east-west direction: a modern version of the cardi and decumani of the Roman tradition.

Las Vegas is different, with the famous Strip, illuminated by huge casino signs, drawing a brilliant and colourful diagonal, which, according to the astronauts, is the most spectacular spot of light on planet Earth.

The colour of lighting varies from one part of the world to another: the Japanese love mercury lamps, which give their cities a greenish colour and make them instantly recognisable. American cities are more yellowish, because they rely on sodium lamps. In Europe, the systems coexist, with the greener downtowns and the newer, yellow-dominated suburban areas. Getting good night photos was not an easy task: the space station's orbital motion of 8 km/s blurs photos taken with the exposure time required by night time panoramas. The astronauts had to be clever at inventing a system to correct the motion of their orbiting home. The first one to get good images was Don Pettit, who achieved a resolution of about 60 m. In order to share his experience with all of us, he used the images to produce a video that offers an extraordinary tour of the world at night and teaches us how to recognize cities based on their illumination.

https://www.youtube.com/watch?v=-RGNhZ292Zg.

Night photos offer a new way to study the urban development of cities around the world. As already mentioned, in addition to tracking population density, lights are also a measure of a country's wealth. The densely populated India is dotted with thousands of small urban agglomerations, but it is a far cry from the general glare of Europe. Income differences are evident when comparing cities that are very close to each other but belong to different states. It happens on the banks of the Rio Grande, which divides El Paso (USA) from Ciudad Juarez (Mexico). The difference is extraordinary: while Ciudad Juarez gives the impression of a shapeless agglomeration, El Paso has a squared structure and, with half the inhabitants, occupies a much larger area, a clear sign of better organization and greater wealth.

Fig. 4.6 Italy by night as seen from the International Space Station, photo by Paolo Nespoli

Browsing through the Nightearth site, one realizes that cities at night offer an extraordinary view. It is sad to note that much of this beauty is due to light (and energy) wasted because it was not directed downwards. All you see is light scattered towards the heavens, absolutely useless for lighting purposes: a foolishly expensive waste that prevents city dwellers from seeing the stars. The fact that this waste has become an unintentional form of art which can be enjoyed only by astronauts, is small consolation, even when the astronauts kindly share it with all of us (Fig. 4.6).

Unavoidable Reflections

Considering the night glow produced by all industrialized nations, it is natural to wonder if there are ways to reduce the energy waste that we are measuring. Obviously, the answer is yes, and the main way is to intervene in regard to street lights and the lighting of public spaces. The rules to follow are simple.

Good lighting must be:

- directional, and only downwards (absolutely avoiding emissions towards the horizon and upwards).
- intelligent, so as to be able to turn on only when needed
- of the right intensity (in order not to waste energy and to avoid the formation of light halos)
- of the right colour (to avoid interfering with the biological rhythms of animals and plants).

However, even applying all of the best practice solutions, we must be aware that it is not possible to completely avoid light reflections towards the sky.

This is demonstrated by an experiment carried out in Tucson, a city in the vanguard of the fight against light pollution that has provided virtuous examples that we will describe later.

To quantify how much of the visible light detected by satellites can be attributed to street lighting, a group of researchers decided to exploit the possibilities offered by the Smart City technology installed in Tucson, which allows for the power of street lights to be varied. During the experiment, conducted over 10 nights, the city administration allowed the brightness of 14,000 of the 20,000 bulbs that illuminate the streets to be varied.

Typically, most of the city's light bulbs are used at 90 percent of their maximum output, and around midnight, their luminosity is reduced to 60 percent. During the experiment, the brightness was reduced by up to 30 percent on some nights and increased by up to 100 percent on others.

According to the authors, the variation in the power of street lights is barely perceptible to people, because human eyes adapt quickly to different levels of brightness. In this regard, I would like to emphasize that the research did not jeopardize public security, since there is no convincing evidence that reducing the brightness of street lighting hampers public safety. This is an important point that is worth considering in our attempts to overcome ancestral fears: flooding public places with light does not decrease the number of crimes.

During the experiment, each night, the team acquired images from <u>Suomi Npp</u>, the newest satellite capable of obtaining high-resolution images of the Earth at night.

The result is surprising: only 20 percent of the light visible from Suomi Npp comes from the streets, while the remaining 80 percent is due to other sources, such as shop windows, stadiums and sports facilities, showing that, even in a city with state-of-the-art street lighting, most of the emissions come from other sources, the contribution of which increases the reflection of light from illuminated objects. Indeed, all illuminated surfaces reflect and, unfortunately, the reflected light is directed upwards. Even asphalt is reflective! Of course, the greater the luminous flux of the lamp, the more light is reflected, so it is a good idea to limit the power of street lamps and switch them off when they are not needed. Intelligent lighting does not decrease public safety, but can contribute to improving the quality of the sky and limiting disturbance to fauna and flora. As we will see later, excessive lighting of the wrong colour is also harmful to our health.

5

How to Measure Darkness

Now that we've gotten an idea of what the Earth looks like at night, it's time to assess the actual relationship between stray light directed upwards (measured by satellites) and the perceived quality of the sky as seen by us Earthlings. As we have come to understand in the search for shooting stars, what really worsens the quality of the sky are the halos caused by the scattering of stray light by molecules and aerosols in the air. We are talking about humidity and dust, naturally present in the atmosphere, and, in general, about everything that is capable of diffusing light, creating the halos of diffuse brightness that amplify and extend the effect of city lighting for tens or hundreds of kilometres.

In order to estimate the scattered light, we start from satellite data and apply a model of light propagation in the atmosphere. The result is cross-checked with many in situ measurements taken with special instruments called Sky Quality Meters (SQM). These are simple, but carefully calibrated, sensors pointed towards the zenith

P. Caraveo, *Saving the Starry Night*, https://doi.org/10.1007/978-3-030-85064-7_5

that can record the light within a given field of view. The result, i.e., the local value of the brightness (or brilliance) of the sky, is expressed in magnitudes per square arcsecond (mag × arcsec^2). This is an astronomical unit of measure that corresponds to an inverse scale of brightness, since a sky of 21 mag × arcsec^2 brightness is darker than one of 20, which is darker than one of 19. The naturally dark sky, free of light pollution, has a brightness of 22 mag × arcsec^2. Since the first step in the battle against light pollution is to assess the extent of the problem, it is extremely important to have a network of SQMs scattered throughout the territory so as to provide continuous monitoring of the situation. The availability of data over long periods makes it possible to follow the evolution of light pollution and exploit particular moments so as to highlight sources of pollution that are not easily traced under normal conditions. In this regard, it's worth mentioning the study done in Veneto by the agency for the environment in collaboration with several departments of the University of Padua during the lockdown period of March, April 2020, when car travel was severely restricted. The researchers focused on studying the first part of the night to see if the decrease in traffic and the shutdown of sports facilities (which could not be used) had an impact on the light pollution recorded at 14 sites across the region. SQMs were used in the centre of Padua, in suburban areas, near the astronomical observatory of Asiago and up in the mountains at different heights. Comparing the measurements obtained during the lockdown (selecting clear nights without a Moon) with the average of the previous years for similar dates and conditions, it was possible to highlight a decrease in the brightness of the sky. The most macroscopic effect was seen in the centre of Padua and at the suburban station, where the SQM recorded a decrease in sky brightness of 20% compared to the 2019 data. The reduction in traffic (and sports lighting) was also felt far from the Veneto plain. In the vicinity of the astronomical observatory at Asiago, where the lighting is regulated so as not to interfere with research activities, the sky's brightness, due to the halo of the lights of the plain, decreased by 10%, while in the highest measuring station (intrinsically darker because it is further away from the illuminated plain), a 5% decrease was recorded.

A comparable decrease in urban light pollution has been measured for Granada (Spain) by scientists from the Astrophysical Institute of Andalusia, who combined images gathered by the Suomi NPP satellite with local SQM data over a period spanning from mid-March to the end of May 2020.

A similar effect has also been recorded in England, where the Campaign to Protect Rural England (CPRE) has launched its annual *Star Count* in the period between February 6 and 14, 2021. As usual, to assess the quality of the sky, CPRE asked *How many stars can you see in Orion?*

Seeing less than 10 stars points to severe light pollution, while counting more than 20 is an indication of a reasonably dark sky, and reaching 30 is a sign of a natural level of darkness, without stray lights.

The data recorded in February 2021 by 8,000 volunteers are the best since 2013. Observations reporting fewer than 10 stars, indicating high light pollution, were down 10% with respect to the previous year, when there were no lockdowns. In addition, 5% of the participants (located mostly in the north of Scotland) counted more than 30 stars, a sign of a very good quality sky.

This is one of the many effects of the pandemic linked to reduced mobility. During the lockdown months of 2020, earth observation satellites measured a 25% decrease in emissions of carbon dioxide, the ultimate greenhouse gas, and as much as a 40% decrease in the production of nitrogen dioxide, which is an indicator of population mobility, because it is produced by car engines. The first images to arrive were from China. Then, in sequence, came the data for the cities, regions and nations that, as the contagion spread, imposed social confinement and a halt to travel. Comparing measurements of nitrogen dioxide concentration recorded in March 2019 with those of 2020 tells a story common to the entire world. A less mobile humanity means that the Earth also vibrates less. Moving crowds, cars, trucks, public transportation, construction sites, and industries produce vibrations that are not individually detectable, but, when added up over large numbers, contribute to the continuous movement of our planet. Data collected by seismographs placed in Brussels, London, Paris, Zagreb, Aberdeen, Los Angeles, Quito and Auckland recorded a decrease in seismic noise between 30 and 50%.

The equation of less-traffic=less-pollution has never seemed more obvious, although it is certainly not new. The study carried out by scientists in Padua, as well as that in Granada, has allowed us to quantify the contribution to light pollution made by the headlights of cars and trucks, a source of light that is not very traceable, but certainly isn't negligible. Astronomers are well aware of this, as they have to learn to drive with their headlights almost completely obscured when they navigate around astronomical observatories at night. It seems impossible, but once we get used to the darkness, our eyes can see even in the dark.

Citizen Science

Taking advantage of the capabilities of the smartphones that we all have in our pockets, we can make our own contributions, even without the use of specific instruments. To get a more precise idea of the quality of the sky at local level, we need the effort of many goodwill observers who are able to join the initiative *Globe at night* (or other similar projects). This is known as Citizen Science, and it is a new way to address scientific issues with the help of all curious and willing citizens. *Globe at night* wants to increase the level of awareness of the impact of light pollution, asking people to measure it with their smartphone. To evaluate the quality of the sky, that is the degree of darkness at local level, you don't need to be an expert nor an amateur astronomer. Similarly to what is done once a year by the Star Count initiative of CPRE, volunteers are asked to take a walk on moonless nights (when the Moon does not illuminate the night because it is physically between the Earth and the Sun), to get used to the darkness and then to look for the constellation chosen by *Globe at night* for that time of year. The purpose of the exercise is to make you aware of how many of the stars, which should be visible to the naked eye, you can actually see. The more stars you see, the better the quality of the sky. Conversely, the higher the stray light, the fewer stars you'll be able to see. Geolocation and the mobile phone clock save you the trouble of entering the coordinates and time of the measurement, at which point you only need to send your observation, which will be loaded into the light pollution database.

In this way, each citizen-scientist will collaborate in measuring the quality of the sky by repeating the exercise as many times as possible, moving to different places during the night. The data collected by thousands of volunteers will be integrated with satellite data and all available information to make a map of the quality of the sky.

It is a simple exercise that, besides giving you a good excuse for a walk, makes it possible to compile a map of light pollution from below. The measurement of sky darkness should be repeated monthly (in fact it is the lunar month, because it is necessary to act in Moonless nights) on the dates that you will find on the site. Perseverance is important. Repeated measurements over time allow to follow the evolution of the situation. The site also contains an interactive map of all the measurements made by volunteers since 2006.

On average, *Globe at night* can count on 100,000 measurements from 115 countries. Of course, the measurements are patchy, with some regions that are heavily measured and some that are almost empty.

Citizen Science to Measure Darkness

Globe at night is asking for the help of as many volunteers as possible to make an accurate map of the quality of the night sky. You don't have to be an expert, just willing to try.

It's all based on an app that you can download from the website http://www.globeatnight.org/webapp/.

What you need to do is look for the right constellation (if you don't know where to start, follow the instructions) and compare the map of a particular region of the sky with what you can see from your vantage point. To make the task easier, the app allows you to vary the depth of the sky chart. You can then appreciate whether you can see up to magnitude 3, 4, 5 or 6. The greater the magnitude value, the better our view of the sky. When there is a good match between the sky chart at a certain magnitude and the sky you see, send the data; the app will add the position and time. If you want, it is possible to add some comments about the weather conditions.

Alternatively, you can use *Loss of the night*, another app that guides you in your search for fainter and fainter stars to allow you to estimate the magnitude limit above your head.

How Light Pollution Changes the Sky

Combining new satellite data with the light propagation model and local measurements, in 2016, Fabio Falchi, Pierantonio Cinzano and Riccardo Furgoni of the Institute of Light Pollution Science and Technology (ISTIL), together with several international collaborators, produced a new map of light pollution "perceived" by observers located all around the globe. Falchi and his collaborators estimated the ratio between the level of darkness measured at each location and that of an uncontaminated sky. This is not an easy task, since the natural level of the sky's background brightness is difficult to establish. We are talking about a quantity that varies depending on the region of the sky being considered, the position of the observer and the quality of the atmosphere.

The results have been summarized in maps of the brightness of the night sky, and it is no coincidence that the nations least affected by light pollution are either the poorest or the least densely populated (Fig. 5.1). Much of Africa is free of parasitic lights, although I fear this is not a choice, but rather a lack of electricity. Rich Australia is equally dark, but, in this case, what matters is the very low population density. You can explore maps online at https://cires.colorado.edu/Artificial-light.

Alternatively, with an analogous method, it is possible to calculate the number of stars visible to the naked eye anywhere on Earth. The task is far from trivial, since, for each place, the limiting magnitude varies in different directions according to the brightness of the sky in each direction, the atmospheric extinction, the distance from the zenith, the height above sea level and the characteristics (and capabilities) of the observer. In order to obtain an estimate of the number of stars visible to the naked eye, the data collected by satellites (which map the emission towards the zenith) are combined with geographical information that takes into account the elevation of a given place and the proximity of other sources of light pollution, the contributions of which are calculated with the atmospheric diffusion model.

Fig. 5.1 The most recent light pollution map produced by Falchi and collaborators in 2016. This time, the map is not based on the light intensity measured by satellite, but rather on the ratio between the perceived level of darkness with respect to the level of natural darkness. Black and grey indicate that the ratio is between 0.01 and 0.04, blue-blue between 0.04 and 0.32, green between 0.33 and 0.64, yellow from 0.64 to 2.56, orange-red from 2.56 to 10.24, pink between 10.24 and 20.5, and white above 20.5 (*From* F. Falchi et al. Science Advances https://advances.sciencemag.org/content/2/6/e1600377)

In such a way, it is possible to know how many stars an average observer will be able to see with the naked eye after adapting to darkness. This method has been applied to Italy, with the results being those shown in the figure (Fig. 5.2).

On the map, the estimated counts range from fewer than 200 stars, for the white zones, to over 1300, for the black ones. Obviously, to be able to see the greatest number of stars, you need to be in dark areas with good atmospheric transparency. These conditions are never present at sea level, where the humidity of the air is a limiting factor. Only the mountains enjoy the best conditions, even if the big cities spread their light for as far as 200 km.

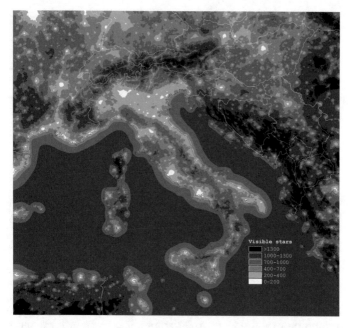

Fig. 5.2 The map of light pollution in Italy. The quality of the sky is measured by the number of stars visible to the naked eye. The fewer stars you can see, the worse the quality of the sky (*Courtesy* F. Falchi from P. Cinzano & F. Falchi Journal of Quantitative Spectroscopy and Radiative, 253, 107059, 2020)

Darkness Ranking

By combining the data from the SQMs, which provide a measure of the brightness of the sky, with the number of stars visible to the naked eye, it is possible to measure darkness and produce a score for sky quality. However, since we often find different units of measure, it is worth looking carefully at the compendium of different ways of measuring sky quality given in Fig. 5.3. On the left, we find the scale with the ratio between the perceived

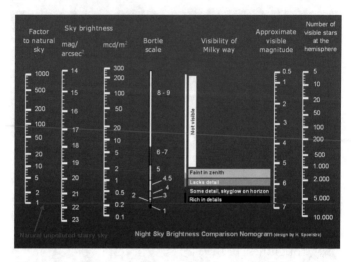

Fig. 5.3 The different ways of quantifying the quality of the dark sky using different units. From the left, the ratio of the background brightness of the sky compared to perfect darkness. The second and third columns show two scales for measuring the brightness of the sky in two different units. The fourth column is the Bortle scale, which gives a synthetic assessment of the quality of the sky, while the fifth column shows the possibility of seeing (more or less well) the Milky Way. The sixth column provides the limiting magnitude achievable by an average observer, while the seventh column gives the number of visible stars for a given limiting magnitude (*Credit* H. Spoelstra)

brightness of the night sky and the natural one. The value 1 points to an uncontaminated sky, which corresponds to the red line crossing the whole figure. In a perfectly dark sky, the amount of stars we can see depends only on the transparency of the atmosphere and the quality of our eyesight. Since not all stars appear equally bright to us, astronomers use magnitudes to describe the intensity of celestial objects, a strange scale that is related to the visual capacity of our eye and, therefore, has a logarithmic trend. To make things even less intuitive, the magnitude scale grows as the intensity decreases: fainter stars have larger magnitudes, while brighter stars have smaller values. A mag 2 star is 2.5 times less intense than a mag 1 star, while a mag 6 star is 100 times fainter than a mag 1 star. Our eye, under ideal conditions, can see stars up to mag 6.6–6.8. Obviously, these values are approximate, because they depend on visual acuity, as well as the experience of the observer. Age is also an important factor, since young people always see better than older persons. These values correspond to a grand total of about 6000 stars visible to the naked eye in the sky (the two columns on the far right of the figure). A dark sky also allows us to appreciate the details of the Milky Way (third column from the right). As the brightness of the sky's background increases, the red line must be moved upwards and the magnitude limit that we can perceive decreases (i.e., we can only see brighter stars), while, at the same time, the number of stars visible to the naked eye decreases as the vision of the Milky Way pales.

In a sky 10 times brighter than the natural value, the number of visible stars has been reduced to fewer than 1,000 and the Milky Way has disappeared. To measure the brightness of the night sky, that is, the luminous flux coming from a certain area of sky, two different units of measurement can be used: the magnitude per arcsec^2 used by astronomers that we have already encountered in the description of the experiment done in Veneto, and the millicandela per m^2, used in lighting engineering. To try to connect this latter unit of measurement with something familiar, let's consider that the brightness of our computer screens is about 300 cd per square metre, equal to one million times the brightness of the dark sky, which is 0.25 millicandela per square metre. In the middle of the graph, we find the Bortle scale, a very straightforward way of

Fig. 5.4 Photographic representation of the Bortle scale, in which 8 represents a sky heavily polluted by the lights of a city centre and 1 is a perfectly dark sky. *Here we see the Orion constellation captured from Bortle Scale Class 1-8 locations. Credit to: Josh Wilson/ Richie Mills/ James Markgraf/ Remco Kemperman/ Robin Lim/ Andrew Wryghte/ Carsten Groinig/ Abhiroop Bhattasali*

quantifying the quality of the sky according to the number of visible stars and the possibility of seeing the Milky Way. Clearly, even the Bortle scale (represented with actual photos in Fig. 5.4) is not at all linear.

6

Biological Rhythms and the Day-Night Cycle

The vast majority of life forms on Earth need the night. The rotation of the Earth has imposed the day-night rhythm, that is, the regular sequence of light and darkness, on all species that have evolved on our planet. This has led to the development of biological clocks that regulate what is called the circadian rhythm through the production of hormones responsible for the sleep cycle and, in general, for our metabolism.

This is particularly true for the animal world, both among vertebrates and invertebrates, including the vast percentage of nocturnal species. During the night, many species go hunting, and a significant portion of migratory birds fly in the dark, perhaps taking advantage of the light of the moon, the only bright celestial body that, through its cycle, provides another natural clock.

The day-night cycle is fundamental to the chlorophyll photosynthesis at the basis of plant life and all animals that feed on plants.

© The Author(s), under exclusive license to Springer Nature Switzerland AG 2021
P. Caraveo, *Saving the Starry Night*,
https://doi.org/10.1007/978-3-030-85064-7_6

Artificial lighting at night (ALAN) now radically alters this natural cycle, and this can cause negative effects on the health of human beings, as well as on the flora and fauna of our planet. The presence of artificial lights causes changes in the habitats of plants and animals, disturbing migrations, reproduction, and the predator-prey relationship, causing accidental deaths in such quantities as to make us fear the extinction of some species. In humans, artificial lighting has a very heavy impact on the production of melatonin, and therefore on the regulation of the circadian rhythm, increasing the risk of hormonal tumours and other serious diseases.

Lighting and Population Health

Light is the most important stimulus for regulating our body's circadian rhythm. As night falls, the pineal gland begins to release melatonin, a substance that is produced only at night and is one of the most studied biomarkers of human physiology. Melatonin regulates the sleep cycle: two hours after the start of production, the body should be asleep. Exposure to light during the night causes an immediate suppression of melatonin production. The effect is all the greater the bluer and more intense the light source, even if it is now clear that human beings are physiologically very sensitive to even low levels of lighting, both indoors and outdoors. Excessive and poorly designed public lighting, in addition to disturbing night vision with dangerous glare side effects, ends up having a negative impact on health, especially in countries where external light is not effectively blocked by curtains or blinds, and thus can enter private spaces. Although it seems hard to believe, the data speak for themselves. In South Korea, a clear correlation has been found between the intensity of light detected by satellites and the sale of sleeping pills. Throughout South Korea, comparing the habits of people who live in areas with different levels of ALAN, it was found that those who live in areas with more light are 20% more likely to sleep less than 6 h, with an average difference of 30 min of sleep between the two samples of the population. Similarly, in the United States, living in highly lit areas has been found to increase the likelihood of sleeping less than 6 h and having poor quality sleep.

In another American study, 29% of people living in brightly lit areas complained of poor sleep quality, while, in darker areas, the number dropped to 16% of respondents. A study on more than 10,000 adolescents showed that sleep disorders, which are linked to anxiety, are more frequent in urban areas than in rural areas. Sleeping in a room that is even dimly lit from the outside causes frequent awakenings and can increase the risk of obesity, high blood pressure, diabetes and depression. Of course, we cannot be certain that all of these adverse consequences arise directly from night-time lighting and the resulting alteration in circadian rhythm. Since ALAN is related to energy consumption and, therefore, to pollution caused by fossil fuels, it is possible that the correlation is real but indirect: people get sick because of pollution, which is, however, much more difficult to measure than light scattered towards the sky.

Returning to the rock solid link between ALAN and melatonin suppression, it should be noted that, in addition to regulating the sleep cycle, melatonin is an effective inhibitor of cancer cell growth. Less melatonin means a greater likelihood of developing certain types of cancer. In a large Harvard University study carried out from 1989 to 2013 on 110,000 women, outdoor lighting was directly correlated with a higher likelihood of developing breast cancer. Women who lived in brighter areas were 14%

more likely to develop breast cancer than those living in much darker areas. A similar effect could also be present for prostate cancer, but the numbers of patients examined in this respect has not been as high as that in the breast cancer study.

On the basis of these studies, it is clear that the lamps we use to illuminate both outdoor and indoor spaces must be designed to minimize the negative consequences for our health. Above all, one should remember that blue frequencies are 5 times more effective in suppressing melatonin (and therefore in disrupting the circadian rhythm) than warmer coloured lamps that do not emit in the blue band. Similar considerations hold true for outdoor lightning. Local administrators should be aware that white LEDs could be harmful and, as such, they should be avoided. Indeed, few countries are already moving ahead to limit the use of blue emitting LEDs on the basis of their adverse effects on the ecosystem as well as on the sky. After severely limiting the blue emission in 2012, Chilean government is considering a total ban in blue emitting LEDs in the Northern part of the country, where the major astronomical observatories are located.

Effects on Plants and Animals

About 30% of vertebrates and 60% of invertebrates are nocturnal species. This means that they have adapted to night conditions by developing sensory capacities adapted to the little light available, relying on the rhythm imposed by the lunar cycle. Artificial lighting can radically alter all of this, causing adverse, and, unfortunately, also lethal, effects on wildlife.

We all know that lights attract insects. In summer, it is not unusual to see clouds of insects gathered around street lamps, especially those that also emit blue and violet light. Although common, it is a local phenomenon, but, in the presence of large concentrations of lights, swarms of insects can assume such proportions as to be able to be detected by weather radar. This happened in Las Vegas, which, between June and July of 2019, experienced an invasion of tens of millions of grasshoppers whose migration path across Nevada was tracked with weather station radar data. Although these mega-swarms are quite rare, it remains true that, even at the local level, the distribution of insects that leave their natural habitat because they are attracted by the lights is altered, and they end up either dying of exhaustion there or being eaten. Indeed, some predators have learned this lesson, patrolling the area around artificial lights where the insects crowd, whilst others, fearful of the light, find it much more difficult to feed. This is what happens to already endangered species such as bats and amphibians, but, all things considered, artificial lights threaten biodiversity because they can negatively affect the entire life cycle, starting with reproduction. In fact, many insects use light to attract a partner and proceed to mating. Indeed, with insect numbers dropping by 80% in some places and no less than 40% of insect species apparently headed toward extinction, the disturbing effects of artificial lightning is starting to be considered as one of the drivers of this apocalypse. In Germany, for instance, it has been computed that 9 million streetlights attract about 1 billion insects a night, a third of which die or are killed by predators. Finding ecologically friendly ALAN is a topic worthy of closer examination. In the Netherlands, a consortium of Universities, industries and non-profit organizations started the Light on Nature project to explore ALAN's effects on local ecosystems. It is a long-term experiment being carried out at seven sets of plots of dark areas. Some plots are lit with lamps of different colours, while others are left dark. Between 2012 and 2016, moth numbers were constant in dark plots, while they decreased by 14% in lighted ones. A similar experiment carried out in a Brazilian tropical forest showed that amber lights attract 60% fewer insect than white ones. Colour sensitivity, however, may vary from species to species. In Pennsylvania, the reactions of fireflies to red, blue and amber lamps were evaluated, and it was determined that red lights interfere the least with courtship, which appears to be significantly disturbed by amber lights that render the females invisible to their potential mates. Although the field of ecologically-friendly lights is still in its infancy, dimmer and redder lights are being tested elsewhere, including at a visitor center in Grand Teton National Park. However, for the sake of insects and ecosystems, preserving dark areas all around the planet is essential.

In the case of migratory birds, artificial lights can be both attractive and repulsive. Many species are attracted to lights and deviate from their routes to get closer, while others carefully avoid them. Attraction can have fatal consequences, as was noted a century ago near lighthouses and fifty years later in regard to airport lights. Birds attracted by lights are far too numerous in urban areas that offer less food than rural areas. Moreover, in cities, there is a high risk of fatal collisions with buildings, especially ones with large glass surfaces. Flocks of birds unquestionably fly at greater heights over cities than over rural areas.

Non-flying animals are also affected by the presence of artificial lights.

Illuminated areas can become *blind ecological spots*, avoided by nocturnal species, which are disturbed by the light, and then by the diurnal ones, because less of their food is available. In the illuminated areas of the Alps, for example, the absence of nocturnal pollinators also results in a decrease in daytime pollinators, who have lost a food source, with negative consequences for the ecology of the entire Alpine environment. The absence of animals is also reflected in the vegetation, which suffers from the reduction of pollination and seed dispersal.

Moving from land to water, it is well known that the presence of artificial lights confuses new-born sea turtles, who, instead of heading towards the ocean, proceed in the wrong direction, with catastrophic consequences. Many aquatic species vary their depth according to the light and can react to artificial lighting by moving closer to the surface or sinking. These shifts alter the predator-prey relationship, because one (or both) is not at the depth where they should be.

Sea creatures are also greatly disturbed by the sound pollution produced by ship propellers and powerful air cannons used for underwater exploration. This is another example of our ability to vary the natural conditions of an ecosystem that is critical to the health of our planet.

From the point of view of plants, artificial lights increase the length of the day, and this can induce an illuminated tree to develop larger leaves that are more sensitive to pollution, because the stomatal pores remain open for longer. Moreover, an illuminated tree loses the perception of the different length of the day in relation to the seasons and, once the summer is over, continues to grow even when it would be better to stop and prepare for winter. If a tree does not lose its leaves in due time, it risks significant damage during winter snowstorms. Of course, not all trees are equally sensitive to artificial light, and this differing susceptibility should be taken into account when choosing plants for brightly lit areas.

However, it remains true that we should avoid illuminating the plants. A brightly lit garden may be pleasing to the eye, but if the plants could talk, they would ask to be left in the dark.

7

In Defence of the Night

Light pollution is a form of alteration of natural conditions. Besides making stars invisible, it affects the biorhythm of human beings, as well as those of animals, insects and plants.

Like any other form of pollution, light pollution has very easily recognizable spatial and temporal characteristics. First, it should be clear that we are talking about medium range pollution. Unlike greenhouse gases, the light emission produced in a certain location does not spread all over the globe. The halo produced by the non-directional emissions of an urban area illuminates a region with a radius of tens or hundreds of kilometres. However, in spite of the limited extension of the halos, it should be kept in mind that photons are impossible to control, because, once emitted, they spread, regardless of political or administrative borders.

© The Author(s), under exclusive license to Springer Nature
Switzerland AG 2021
P. Caraveo, *Saving the Starry Night*,
https://doi.org/10.1007/978-3-030-85064-7_7

So, the level of deterioration in the quality of the sky above us depends very significantly on what is happening in the surrounding areas.

However, unlike the emission of greenhouse gases, which linger in the atmosphere for a very long time, photons have a short life, and if light sources are turned off, the situation improves immediately. This is good news, demonstrating that the problem, although serious and widespread, is not irreversible and could be kept under control by adopting some "reasonable" countermeasures and raising public awareness. Non-profit associations, operating both at the global and regional levels, are fighting for the protection of the dark skies and the night environment. Thus, we should be grateful to the International Dark sky Association (IDA) as well as to smaller national groups. In Italy, the bulk of the work is carried out by Cielo Buio, together with the Institute of Science and Technology of Light Pollution.

Limiting the amount of stray light is relatively simple: to begin with, one should use lamps that illuminate the ground and not the sky. As obvious as this may seem, this is almost never the case: a sizable fraction of energy is directed (and wasted) upwards or towards the horizon. The difference between a good lamp and a bad one is not a matter of price, but simply of design and attention. Lamps in the shape of a luminous sphere, for example, are the worst imaginable, because they illuminate everything but the street below. With directional lamps, the power, used to illuminate a narrow and well-defined region, can be usefully reduced, decreasing energy consumption and reducing the undue illumination of the sky, which, unfortunately, cannot be completely eliminated, because, even in the best possible conditions, it is never possible to eliminate the light reflected from the illuminated surfaces.

Controlling the quality of stray light is less straightforward, but certainly not impossible. In this context, it is worth reminding how the Mount Palomar and Kitt Peak Observatories were saved.

Let's start with Mount Palomar Observatory. Although it was built before the Second World War, it started operating, with its 5 m mirror, in 1948. At the time of site selection, the region around the city of San Diego (only 70 km away) had about 210,000 inhabitants; fifty years later, in 1980, the population had exceeded 1,800,000, mostly settled in new residential areas built in rural regions around the Observatory. In 1960, the problem of light pollution was already an issue at the Observatory: the sky was about twice as bright as in Las Campanas (in the Chilean Andes), and in the spectra of celestial objects, there were lines of mercury from street lamps.

Starlight and Street Lighting

To understand the nature of celestial bodies, astronomers have a very powerful method of investigation: spectroscopy. Light contains a lot of information about the object that produced it. If we disperse light with a prism (as Newton did centuries ago), we obtain a spectrum. What appears to us as white light is broken down into the colours of the rainbow. Observing this with appropriate instrumentation, we realise that indentations, holes, appear in the colours. Astronomers call them absorption lines and use them to understand the composition of celestial objects. Hydrogen, helium, carbon, oxygen, nitrogen, and then gradually all of the elements, are responsible for lines at certain wavelengths. By recognizing the line, we recognize the element. It is clear, therefore, that if we want to understand what stars are made of, we must not pollute the spectrum with too many parasitic lines produced by the lamps used for street lighting. Astronomers have learned to live with a few lines that they are able to "filter out," but they cannot work miracles.

Since lighting is considered an essential asset, it was a question of trying to reconcile the needs of astronomers with those of the rest of the world, remembering that the problem of light pollution is twofold: what matters is not only the amount of stray light, but also its spectral characteristics. It is evident that, if street lighting uses lamps that cover a wide band of the spectrum, introducing parasite lines here and there, astronomical research is highly penalized. Hence the need to limit the region of the spectrum to be shared as much as possible by moving towards use of monochromatic lamps. Low-pressure sodium lamps combine "spectral" advantages with economic ones. First of all, they produce only a few lines, the most intense of which is in yellow, the best wavelength for our eyes; moreover, they are the most efficient solution, as their emission is the highest per watt of energy absorbed. Monochromatic yellow lighting is also considered the safest for drivers, because it does not dazzle and improves contrast perception, moreover, it is less scattered by fog and rain.

All these advantages come at a small price: precisely because they are monochromatically "yellow," these lamps have the unpleasant effect of giving an irremediably yellowish appearance to things and people. Much more pleasant is the light of a mercury lamp, which, unfortunately, in addition to being less efficient, has other substantial disadvantages. The mercury emission lines are in the blue green segment (where our eyes are less sensitive), with a dangerous appendix in the ultraviolet. Thus, to avoid the production of harmful UV, other gases are added in order to absorb and re-emit such lines at other wavelengths, transforming a source of lines in a continuous spectrum, with the resulting astronomical damage mentioned above. Indeed, aesthetic considerations have, for some time, frozen the projects concerning light pollution control in the region around the Observatory. However, in 1981, the

County of San Diego passed an ordinance on the use of low-pressure sodium lamps whenever good colour rendering is not mandatory, typically in the lighting of roads, parking lots, and sports facilities. In addition, limits have been placed on the use of conventional lighting, whether for commercial and social purposes or for decorative ones, mandating that it must be compulsorily switched off at 11 p.m. These measures have been very effective in limiting the damage caused by the coexistence of a large observatory with a growing urban settlement. Similar stories have happened or are happening in regard to all large Observatories located in (or near) densely populated areas.

Mount Hopkins and Kitt Peak, for example, operate very satisfactorily, despite their relative proximity to Tucson, another booming urban area. Here again, the trump card has been campaigns to raise awareness among local officials and the public, resulting in strict (but reasonable) ordinances on the types of lamps to be used for private and public lighting. In Tucson County, it is illegal to use mercury vapour lamps for street lighting, and tens of thousands of lamps have been replaced with low-pressure sodium lamps. As a result, Tucson has a sky that is at least 3 times darker than it would be without control, and its residents, while having sufficient lighting to live normally, can enjoy the view of the Milky Way on summer evenings.

Unfortunately, not all local administrations have the same sensitivity to the problems of light pollution. In any case, the first step is always to be aware of the existence of the problem and to wonder which lamps can provide the best cost-benefit ratio without scattering the light upwards and without emitting in the blue-violet part of the spectrum, which is the one that causes the most harm to both astronomy and our circadian rhythm. If monochromatic yellow light is not deemed appropriate in urban areas, there are reasonable alternatives with high-pressure sodium lamps that, although less efficient, have a better colour rendering.

However, while these considerations continue to hold true, the problem of lighting has been revolutionized by LEDs (Light Emitting Diode), light sources that combine brightness with remarkably low power consumption. What more could you ask? Just a bit of good sense. Since LEDs are cheap and efficient, it's easy to install too many lamps, resulting in a far too bright ALAN that increases (by a lot) the light halo around inhabited areas and, in fact, worsens light pollution.

Moreover, many of the LEDs on the market produce "white" emission, that is, they have an important blue-violet component in their spectrum, precisely the wavelengths that are more effectively diffused in the atmosphere and more efficient for disturbing the melatonin cycle in humans and many other species. These considerations should induce public administrators not to use white LEDs for street lighting, and each of us should think about them when choosing the light sources to illuminate our homes. Why use a potentially harmful product? Fortunately, lamp manufacturers offer a wide choice, and it's easy to find LEDs of warmer shades without emissions in the blue-violet part of the spectrum.

The four Pillars for a friendly and efficient light scheme are:

The **right** light, at the **right** place, of the **right** brightness for the **right** duration.

Darkness Must Be Protected

Legislation to protect the sky's darkness varies greatly from region to region and from state to state. To date, only a handful of states (Chile, Croatia, the Czech Republic,

France, Italy, Slovenia and Spain) have adopted laws at the regional or national level in order to limit the light pollution produced by public lighting, but I hope that every local and national body will address the problem following the indications of the International Dark sky Association.

In Italy, for example, thanks to the work of Cielo Buio and the Italian section of the International Dark sky Association, all but two regions have adopted laws to control light pollution. The first to take this step was the Veneto region, which regulated the matter in 1997, and then returned to the subject in 2009, with more restrictive requirements that impose the zeroing out of any upward emission. After all, in Asiago, a town in the Veneto region, there is the Observatory of the University of Padua, with the largest telescope on Italian soil.

Additionally, Piedmont, Liguria, Trentino Alto Adige, Friuli Venezia Giulia, Emilia Romagna, Marche, Umbria, Abruzzo, Molise, Puglia and Sardinia have laws that, following the example of Lombardy, impose the zero emission upwards, together with a reasonable curfew for unnecessary lights; Tuscany, Lazio and Campania, instead, admit exceptions that are more extensive for Valle d'Aosta and Basilicata. We don't know what light pollution would be like in Italy if all of these meritorious initiatives had not taken place, but we do know for sure that the quality of the sky in the zero emission regions has not worsened during recent years and, considering global light pollution growth, this is an important result all by itself.

In most cases, simple measures and good will are sufficient to significantly reduce light pollution and energy consumption. The energy savings are such that they quickly cover the costs of buying and installing the new lamps. Moreover, the interest aroused in the public has also induced the industries to produce better lamps and to study more efficient systems for the lighting of road signs and sports facilities. Ultimately, the International Dark sky Association proves that the only battles lost are those that are not fought! Unfortunately, victories have to be continually renegotiated, taking new technologies into account.

Good and Bad Examples

However, it remains true that attention and good will are the best weapons to avoid further worsening the state of light pollution.

Turning off lighting on monuments when it is not necessary is a simple but effective way to take action. As of July 1, 2013, France has decided to limit the night-time lighting of all non-residential buildings, whether they are monuments, office buildings, shops or advertising signs. In Paris, for example, all monuments' lights are turned off after 1a.m., when very few people are out for sightseeing. The 6-h switch-off, from 1 to 7 a.m., does not affect private homes, nor does it affect the lighting of streets, stations and public utilities. Although beneficial for limiting light pollution, this decision was not an astronomical initiative. The purpose of turning off unnecessary lights is to save on electricity costs, to limit damage to the ecosystem and to cut the amount of carbon dioxide released into the atmosphere by the combustion of oil, one of the ways to produce electricity.

Fans of the (beautiful) 20,000 LED lighting of the Eiffel Tower should not feel deprived. The show is turned off late at night.

But it is not only an economic issue, because, as we have seen, too much night lighting disturbs the rhythm of almost all nocturnal animals (including humans) and can affect the migration of birds that fly in the dark by orienting themselves with the stars.

In fact, the six hours of darkness is only the minimum amount required; lights can be switched off earlier in such places as empty office buildings. The same goes for shops, which have been asked to turn off their windows and signs one hour after closing. This idea did not at all please shop-keepers, who bargained for a period of darkness limited to late at night, fearing repercussions on tourism.

Not all examples of experiments with lighting are virtuous, and thus it is also worth mentioning those that should be carefully avoided.

While Paris is trying to curb unnecessary lighting, the administrators of a few resorts in Switzerland have decided to "daylight" some of their most iconic mountains. In one fell swoop, with considerable energy expenditure, they managed to turn off the stars and disturb all the nocturnal animals. Here's an example NOT to be followed (Fig. 7.1).

Fig. 7.1 A stark example of a very bad idea: undue illumination of the Swiss mountains. The photo "of the Aiguille de la TSA irradiated with artificial light" is from some time ago. I hope that, in the meantime, the widespread protests prompted by the publication of this photo have convinced the local administrators to stop this wasteful and useless practise (*Credit* Christopher Blott for Keystone)

Let's never forget that our individual choices have repercussions on a much more general level. Let's think about it! Sometimes all it takes is a little attention to make a difference and avoid doing harm to the ecosystem while also switching off a few stars.

Let's mind the night (and our health) when we choose the lighting for our gardens, or our streets.

Let's learn to respect the night. The sky is an important part of our culture, and it is our duty to try to preserve it while minimizing the ALAN disturbance on the ecosystem.

8

Finding the Dark

Celestial phenomena attract crowds of enthusiasts. Total eclipses of the Sun are a real magnet, capable of attracting thousands of people who will not hesitate to travel to the other side of the planet to enjoy the view of the Sun going dark. Totality is a fascinating moment, full of ancient suggestions, that makes us unique in the solar system. Witnessing the darkening of the Sun is, in fact, only possible for us Earthlings, thanks to a fortuitous combination of the size and distance of both the Sun and Moon, as seen from Earth. Although the Moon's diameter is 400 times smaller than the Sun's, the distance between the Earth and the Sun is about 400 times greater than the distance between the Earth and the Moon. It is this coincidence, unique in the solar system, that makes it possible for the small Moon to obscure the Sun. Eclipses can happen only during the New Moon, when the Moon is between the Sun and Earth and cannot be seen at night. Not all New Moons, which occur about once a month, produce eclipses, of course. It depends

P. Caraveo, *Saving the Starry Night*,
https://doi.org/10.1007/978-3-030-85064-7_8

on the 5° inclination of the Moon's orbital plane relative to the plane of the Sun's apparent orbit, which astronomers call the ecliptic, precisely because it is the plane where eclipses occur. To allow for superposition between the two celestial bodies, the Moon's orbit must cross the ecliptic at the right time. This is what makes eclipses a fairly rare and always fascinating phenomenon. The record for public interest was reached during the Great American Eclipse that crossed the entire United States on the morning of August 21, 2017. In addition to the 12 million people residing within the totality zone, another 47 million Americans were able to reach it with a car trip of about two hours and, extending the travel time to one day, the total number of potential observers reached the record figure of 100 million. In addition, many astronomical tourists organized their trips by choosing mostly anonymous locations that, however, had the advantage of being within the totality zone.

Less dramatic, but no less impressive, is the sight offered by the lunar eclipse that occurs when the Sun, Earth and Moon are aligned (in that order). The Earth blocks the Sun's light and casts its shadow on the Moon, which, no longer illuminated, should plunge into darkness. In fact, during a lunar eclipse, the full Moon transits for about an hour within the Earth's cone of penumbra and its brightness progressively fades, after which it enters the cone of Earth's shadow and the illuminated part of the Moon diminishes as the minutes go by. However, at a moment when darkness should rule, the Moon turns red! This is an effect resulting from the Earth's atmosphere, which absorbs all the colours of the Sun's light except red. The Moon is thus illuminated by a red light and shines with a characteristic reddish shade. The phenomenon lasts for as long as it takes the Moon to pass through the cone of shadow, let's say roughly an hour, but the duration also depends on the distance of the Moon from the Earth at the moment of the eclipse (Fig. 8.1).

Fig. 8.1 Red moon (Luna Rossa) over Italy, July 20, 2019 (*Credit* INAF)

A lunar eclipse can only happen at night, and it's definitely more "democratic" than the Sun's eclipse, since you can admire it from the entire terrestrial hemisphere, which is in the dark (and has the Moon above the horizon, of course). You don't have to face long journeys to admire the Red Moon, you just need to find a place not too affected by light pollution, the ideal being a Dark Sky Park or, even better, a Dark Sky Oasis.

The Oases of Darkness

Looking again at the map of upward light over Europe (Fig. 4.4), we realise how difficult is to find a pristine sky. Only a few areas in Scotland and in the far north of the Scandinavian peninsula enjoy truly dark skies. In the

United States, which is, on average, less populated than Europe, dark areas are more numerous, covering high altitude and less inhabited areas.

Where should a person who wants to enjoy the sight of the starry sky go? To answer this question, we can consult the list of darkness oases compiled by the IDA in collaboration with other associations, such as the Starlight Foundation and the Royal Astronomical Society of Canada. The list contains 223 oases of darkness classified into six categories, according to a scheme drawn up by the International Union for the Conservation of Nature, which includes:

- Dark Sky astronomical sites (15), characterized by excellent conditions, with transparent atmospheres and total darkness, generally in remote places at high altitude, suitable for professional astronomical observations
- Dark Sky parks (114), protected natural areas and uninhabited nature reserves, thus free of artificial lights, where the defense of the dark goes hand in hand with the protection of the entire ecosystem
- Dark Sky heritage sites (9), protected areas that are also of historical interest
- Dark Sky outreach sites (25), which can be urban, suburban or rural areas where astronomy outreach activities are carried out
- Dark Sky reserves (21), where a protected area is embedded within a community that collaborates towards sustainable development respecting the dark
- Dark Sky communities (39), entire towns where the inhabitants agree to apply the rules for limiting light pollution.

These 223 oases of darkness cover just 0.14% of the Earth's land area, a very small fraction of our planet.

However, it is very important that they exist, because they are examples of what can be achieved through collaboration with a public that agrees to follow the rules of responsible lighting with the aim of protecting the dark sky and biodiversity, as well as adding value to their area. In fact, if darkness is hard to find, it means that a dark place could have economic value, attracting tourists who travel precisely in search of darkness.

Let us not forget, moreover, that protecting the night also has important ecological value, because a naturally dark environment becomes a refuge for wild animals that are disturbed, sometimes severely, by artificial lights. Darkness, together with the lunar cycle, is an indispensable element for the health and balance of all the beings that call our planet home.

Astronomers Conquer Mountain Tops

By now, it should be clear why astronomers build their most powerful observatories in remote and secluded places.

Since "population" rhymes with "illumination," both literally and figuratively, astronomers must flee populated areas. In addition, one must look for the right geographic setting that ensures a high number of clear nights and excellent sky transparency. In general, the quality of the sky is better at high altitudes, especially in some locations that are always above the thermal inversion layer, that is, above the cloud layer. This is a phenomenon that can be seen on the summits of some islands, where the mountain peaks dominate over a sea of clouds formed by the condensation of humidity from the oceans that surround them. This happens in the Canary Islands, where all European nations have built their observatories, and in

Hawaii, where the mountaintops of many islands serve as astronomical heavens (Figs. 8.2 and 8.3).

The legendary Mauna Kea is so crowded with telescopes that the native Hawaiians are fighting to prevent a new large 30 m diameter telescope from being built there, because they fear that the colossal structure will disturb the spirits of their ancestors, who, according to tradition, live on the high peaks. I honestly don't think that an astronomical observatory, however large, would disturb the spirits of the ancestors, who certainly revered the sky as much as modern astronomers do. What the native Hawaiians fear is rather a kind of cultural colonization aimed at exploiting the extraordinary quality of the sky without adequate cultural and scientific return to the archipelago. I believe, and hope, that a solution will be found that is in the best interest of both Hawaii's traditions and astronomical research.

However, both the Canary Islands and Hawaii are in the Northern Hemisphere. In order to observe the southern sky, both European and American scientists have

Fig. 8.2 Photo of the telescopes on top of the Mauna Kea volcano (*Credit* Alan L from Flickr)

Fig. 8.3 Photo of the INAF Galileo National Telescope on the crest of the caldera of the Taburiente volcano on the island of La Palma in the Canary Islands above a sea of clouds (*Credit* INAF)

chosen remote locations in the Chilean Andes. These are mountains that, while in desert areas, are relatively close to the Pacific Ocean, and thus offer the same advantages as the islands, because they are above the oceanic cloud layer and have excellent quality skies. And when we say desert, we are not exaggerating at all.

The European Southern Observatory (ESO), the international organization that was formed to provide European nations with access to the southern hemisphere, has built its Very Large Telescope (VLT) in the Atacama Desert, where it rains less than in the Sahara. The landscape is Martian, but the quality of the sky is among the best in the world. The land was donated by the Chilean government, but, to protect itself from possible intrusions, ESO bought the mining rights to the area, which must remain uninhabited, free of industrial settlements and completely dark (Fig. 8.4).

Fig. 8.4 The four 8 m diameter telescopes that make up the VLT shine in the rays of the setting sun atop Cerro Paranal in the Atacama desert (*Credit* ESO)

The efforts to defend the extraordinary quality of the sky benefit today's astronomers as well as future ones. In order to maintain world leadership in European astronomy, a few kilometres away from the Very Large Telescope, ESO is building the Extremely Large Telescope (ELT), a giant with a 39 m-diameter segmented mirror. This is a technological jewel, with an important Italian contribution, since the companies that won the contract for the construction of the mammoth, but very manoeuvrable, structure are Italian, as will be the instrument that will guarantee the telescope a very sharp view thanks to the technology of active optics developed by the National Institute of Astrophysics to cancel the effects due to the instability of the atmosphere.

The ELT was conceived to study the first stars to ignite in the Universe, or to search for biomarkers, i.e., gases that can be linked to the presence of some form of life in the atmosphere of extrasolar planets. However, like any well designed, new and powerful instrument, it will be ready to take on new challenges, to study new problems that will open up in the years to come.

Obviously, the extraordinary quality of the skies of the Chilean Andes is also a magnet for American astronomers, who have built large observatories and are in the process of building new ones. Among these, I would like to mention the large observatory that is under construction on Cerro Pachon. The feature that makes it unique within the astronomical landscape is its huge field of view, designed to allow it to cover vast portions of the sky. When it was born, the project was called the Large Synoptic Survey telescope (LSST), but in June 2019, the chairwoman of the House Committee on Science, Space, & Technology of the US Congress presented a proposal to change the name of the telescope under construction into the Vera Rubin Survey Telescope. The proposal, later approved by the U.S. House, offers an opportunity to remember a great astronomer, who, in addition to having made fundamental contributions to research, was also an extraordinary advocate for opening the discipline to women.

The choice to dedicate the LSST to Vera Rubin is not at all random, because the telescope was designed to study dark matter and the even more mysterious dark energy that dominate our universe.

With its large field of view, which will be covered by a camera yielding 3.2 gigapixel images, the telescope will map the distribution of billions of galaxies, an avalanche of data that will help us better understand the dark side of the cosmos, a surprisingly big fraction of our Universe discovered by Vera Rubin half a century ago.

Back in the 1960s, life for women who wanted to dedicate themselves to science was not easy. Vera was the mother of four children and had to fight against all kinds of prejudice. In 1965, she sent her first application for an observation at the telescope at Mount Palomar, which was then closed to women because there were no suitable bathrooms. Rubin's solution is famous: she cut out a stylised figurine of a woman and stuck it on the bathroom door. The figurine did not survive for long, but the taboo was overcome, and Rubin was the first female astronomer to be permitted to use the Hale telescope at Mount Palomar. In truth, Margaret Burbidge had been there before her, but Burbidge had applied on behalf of her husband, Geoffrey Burbidge, a famous theoretical astrophysicist. Everyone knew that he would not be the one to observe, but appearances were safe.

After spending years measuring the anomalous motion of galaxies and their distribution (obtaining results that raised a lot of criticism and would be accepted only decades later), Rubin decided that she was tired of being involved in controversial topics and chose to study the rotation curve of galaxies, a subject that, at first glance, seemed to her to be peaceful. But she was wrong!

The motion of stars within galaxies was different from what might have been expected from Kepler's laws. Each star moves by "responding" to the mass contained within its orbit. As we move away from the center, the stars should slow down, as happens with the planets in our solar system. Instead, Rubin measured flat rotation curves; what she found was that the stars didn't slow down at all. In other words, the stars were moving too fast and should have gone their own way, no longer being held by the gravitational pull of their galaxy. Thus, since the galaxies seemed to be stable, one had to imagine that they contained a far greater amount of mass than was visible in the form of stars, gas, and dust, a discrepancy that had already been noted by Frank Zwicky for the Coma galaxy cluster.

Thus, Rubin became the mother of galactic dark matter, one of the most extraordinary achievements of astrophysics in the last century. Thanks to her results, we now know that 90% of the matter in our galaxy (and all others) does not emit radiation that can be measured by our instruments. We know very little of this pervasive and obscure mass, and we have no clues as to its constituents.

Vera Rubin didn't receive the Nobel Prize, which she would have deserved, but she is the first woman astronomer to be honored with her own telescope and, as we'll see, the Vera Rubin Telescope will have to fight against new challenges that risk ruining its huge and beautiful images.

9

Radio Waves also Suffer from (Electromagnetic) Pollution

So far, we have talked about optical astronomy, but we cannot forget about radio astronomy, which, even though it operates at different frequencies from the optical, has similar problems. The radio emission of celestial bodies has wavelengths that are measured in mm, cm, m up to tens of kilometres, not in fractions of a thousandth of a mm like the optical ones. Shorter wavelengths correspond to higher frequencies, because the frequency of a wave is proportional to the inverse of the wavelength.

It is worth remembering that the development of radio astronomy from the ground is possible thanks to the transparency of the atmosphere to a large part of the immense frequency range covered by radio waves, which, as we see in the Fig. 9.1, can have lengths that vary from kilometres to mm. We have already mentioned the courtesy that the atmosphere offers only to radio and optical photons. The rest of the electromagnetic spectrum is absorbed, and thus it vanishes. For this reason, when talking about light

© The Author(s), under exclusive license to Springer Nature Switzerland AG 2021
P. Caraveo, *Saving the Starry Night*,
https://doi.org/10.1007/978-3-030-85064-7_9

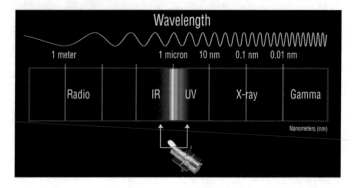

Fig. 9.1 Representation of the electromagnetic spectrum that highlights the wavelengths of the different radiations studied by astronomers. Only the radiations that fall within the radio and optical domains reach the ground; all the others are absorbed by the atmosphere and must be studied with instruments aboard satellites. Observing above the atmosphere has advantages in the optical domain as well, as we learned from the wonderful images from the Hubble Space Telescope, but the vast majority of optical observations are done with instruments on the ground (*Credit* NASA). https://www.nasa.gov/sites/default/files/styles/full_width/public/thumbnails/image/hst-electromagnetic-spectrum-2.gif?itok=k62ouNpK

pollution, and, more generally, electromagnetic pollution, we deal only with radio and optical astronomy. This does not mean that the other astronomies are less interesting, they simply cannot be done from the ground, and this frees them from the problems of electromagnetic pollution (even if it is not easy to build and operate instruments outside of the Earth's atmosphere).

Radio waves of celestial origin were discovered by chance in 1932, by Karl Jansky, a young engineer at Bell Laboratories. He was trying to understand the origin of the noise that disturbed voice communications across the Atlantic when short waves were used. Using a large directional antenna, Jansky realized that the disturbance was at a maximum once a day and thought it was related to the Sun. A closer look at the timing of the signal showed that the interval between one emission maximum and the next was 23 h and 56 min, what astronomers call a sidereal day. It was clear evidence that the source was celestial. He then looked at a map of the sky and found that the source of the emission was the central part of the Milky Way in the constellation Sagittarius. What was responsible for the radio emission? Jansky thought it was due to interstellar gas and dust, but he was wrong. Now, we know that the emission is produced by electrons that lose energy as they spiral around the magnetic fields that permeate the entire galaxy. After announcing his discovery in 1933, Jansky hoped to further study the radio emission of the Milky Way, but Bell Laboratories entrusted him with another project that took him away from radio astronomy. However, his work has not been forgotten: the radio flux of celestial sources is measured in a unit called the jansky (Jy).

The study of the spectra of stars had taught us that, to get useful astronomical information besides the continuous emission, spectral lines are essential. Only by measuring the wavelength and the amplitude of the lines can astronomers recognize the elements, estimate their temperature and, if the lines are seen at wavelengths different from the canonical ones, their speed. Lines would have been of great help for radio astronomers who measured a radio flux from our galaxy but had no idea where it was coming from. If there had been a line, it would have been possible to estimate the speed of the emitting regions, and thus model their distance from the center of the galaxy. In the spring of 1944 (with Holland occupied by Nazis), the great Dutch astronomer Jan Oort posed the problem to the young Hendrik van de Hulst, who focused on hydrogen, the most abundant element in our galaxy and, indeed, in the Universe. Clearly, it was not advisable to look at the quantum jumps of electrons from one orbital to another, those that are responsible for the optical lines, with energy much higher than the radio frequencies. Van de Hulst concentrated on the spin of the only electron of the hydrogen atom. Since the proton also has its own spin, two configurations are possible for the hydrogen atom with parallel or antiparallel spins. This last configuration is the one of lower energy, so it is the one preferred in nature. However, it may happen that, following collisions with other particles, the atom is excited and acquires a small amount of extra energy that changes the spin of the electron, which will tend to return to the lowest energy state by emitting radiation at a wavelength of 21 cm. This is a very unlikely hyperfine transition, since, for any single atom, it happens once every 10 million years. However, on a galactic scale, the low probability of the transition is compensated for by the very large number of hydrogen atoms present. Van de Hulst published his results in 1945, and the 21 cm line was measured in the laboratory in 1951, becoming a workhorse for radio astronomers, who could use it to map the spiral structure of our galaxy, something we cannot see, because the optical photons are absorbed by interstellar dust, which, instead, is transparent to radio waves.

Single and Multiple Radio Telescopes

The development of radars for war purposes proved to be very useful in improving radio astronomy techniques based both on fixed instruments, which observe the sky as it passes over them, and on steerable telescopes, capable of pointing in any direction. Their dimensions became larger and larger so as to offer a greater collecting surface, but also a better angular resolution, a quantity that is inversely proportional to the diameter of the telescope. In other words, a larger telescope offers two advantages: it has a sharper view and reveals fainter sources.

Fascinated by the possibilities offered by the 21 cm line, Jan Oort proposed building a 25 m diameter telescope, which was inaugurated in Dwingeloo, Holland, in 1956. The following year, it was the turn of the 76 m diameter Jodrell Bank telescope in Manchester, while, in 1961, the Parkes telescope in Australia, with its 64 m antenna, became operational.

In the United States, radio astronomy activities were coordinated by the National Radio Astronomy Observatory (NRAO), founded in 1957. The first antennas, both transit and steerable, were built in Green Bank, Virginia, where, in 1962, the 92 m diameter steerable telescope came into operation. It was used until 1988, when it collapsed and was replaced by a new 100 m diameter telescope, which challenged the primacy of another 100 m radio telescope built in 1971, in Effelsberg, Germany.

In matters of size, however, no one could rival the Arecibo Transit Radio Telescope, featuring a 305 m diameter "dish" lying above a natural depression in a forest on the island of Puerto Rico (Fig. 9.2). It came into operation in 1963, and, for more than 50 years, was the largest in the world, until it was overtaken by China's FAST telescope in 2016.

The gigantic structure had arisen from the military's interest in tracking nuclear warheads in the upper atmosphere or in low orbit. For this reason, it was built so that it could be used as a receiver (like all radio telescopes), but also as a transmitter, a unique capability that allowed it to study many asteroids that passed close to Earth. It was its planetary radar that obtained, in 1967, the first important result, with the measurement of Mercury's rotation period.

Fig. 9.2 Aerial photo of the Arecibo telescope (*Credit* NAIC - Arecibo Observatory, a facility of the NSF)

Arecibo has made fundamental contributions to astrophysics, collecting data that led Russell Hulse and Joseph Taylor to be awarded the Nobel Prize in 1993, sending the first message to aliens in 1974, and also managing to enter the collective imagination thanks to the film *Contact*.

It was an extraordinary and iconic telescope that, after enduring violent tropical hurricanes, earthquakes, and repeated funding cuts, surrendered to the wear and tear of time in December 2020, with a crash that was recorded by seismographs. The government of the island of Puerto Rico has decided that it will be rebuilt.

In parallel, the technique of synthesis aperture was born, thanks to which several radio telescopes can be operated as one so as to virtually increase the size of the collecting surface and improve the angular resolution of the images produced.

This technique was born in England at the end of the 1940s, and was applied everywhere in the world, with observatories formed by many antennas designed to be moved so as to vary the distance between them. The most spectacular is ALMA (for Atacama Large Millimeter Array), at an altitude of 5,000 m, on the Puna de Atacama plateau, in Chile. It is an international project based on contributions by ESO, the United States, Canada, Japan, South Korea, Taiwan and Chile. It features 66 antennas with diameters of 12 and 7 m that can move over the desert plateau for distances from 150 m to 16 km (Fig. 9.3).

It's a very successful idea, and is the basis for the Square Kilometer Array (SKA), an ambitious project that represents the future of radio astronomy. It will consist of thousands of antennas organized in two observatories separated by the Indian Ocean, one in Australia and one in South Africa. The choice of the sites was based on the data of the electromagnetic noise produced by our planet measured from orbit. Indeed, west Australia and South Africa enjoy

Fig. 9.3   Aerial view of ALMA with its 66 antennas (*Credit* ESO)

a unique electromagnetic cleanliness. These are vast desert regions, very sparsely populated.

At first glance, the two observatories will have very little in common, since they'll host arrays of different-looking antennas optimized to pick up radio signals at different wavelengths. In South Africa, there will be an array of 197 radio telescopes (the first 64 will have a 13.5 m diameter, while the remaining 133 will have a diameter of 15 m), while, in Australia, 512 arrays of 256 omnidirectional antennas will produce data that will be combined via software (Fig. 9.4).

In both cases, however, the antennas will cover vast spaces and will be arranged according to well-defined geometries. Despite being on two continents thousands of kilometres apart, the sites have similar latitude, so they share the same sky. Once completed, the instruments will be able to function as a single observatory and will have performances equivalent to that of an antenna with a diameter of

Fig. 9.4 Composite image of the SKA telescopes, blending real hardware already on site with artist's impressions. From left: artist's impression of the future SKA dishes blend into the existing precursor MeerKAT telescope dishes in South Africa. From right: artist's impression of the future SKA-Low stations blends into the existing AAVS2.0 prototype station in Western Australia (*Credit* SKAO)

Fig. 9.5 Inaugural image produced by the array of antennas that make up MeerKAT, the precursor to the SKA observatory (*Credit* SARAO)

one square kilometre (as the name implies). Construction of the two observational sites will proceed in stages, beginning with precursors that are designed to represent a fraction of the SKA project's collecting area. MeerKAT, consisting of 64 13.5 m diameter telescopes, was inaugurated in South Africa in 2018. The SKA precursor started to demonstrate its capabilities with a spectacular image of the center of our galaxy at the frequency of the 21 cm hydrogen line (Fig. 9.5).

The Shadow of the Black Hole

By synchronizing telescopes located in different parts of the globe (and also in space), the distance between the antennas is maximized, thus achieving the best possible angular resolution. This is called VLBI (Very Long Baseline Interferometry), a difficult technique that requires worldwide chronometric coordination, but allows us to reach angular resolutions of less than one thousandth of an arc second.

To capture the shadow of the black hole of a few billion solar masses at the center of M87, it was necessary to push the technique to the limits of the possible. M87, a giant galaxy at the center of the Virgo cluster, has a powerful radio jet, evidence of the activity of its central black hole, the mass of which has been estimated at about 6.5 billion solar masses. Although black holes don't emit anything, because no type of radiation can escape their gravitational attraction, they can be very bright celestial sources, thanks to the matter that spirals around them, forming accretion disks. These are large gravitational waiting rooms, where gas thickens and heats up while waiting for its time to cross the event horizon and disappear into the black hole. Optical astronomers and those working with X-ray instruments in orbit are used to studying these monsters indirectly. They see the emission produced by matter spiralling around the black hole within the accretion disk and the powerful jets where particles are accelerated to near light speed. It had never been possible to see what happens in the vicinity of the black hole, where the event horizon divides the inside from the outside, for the very good reason that, on a cosmic scale, it is a very small region.

Considering the relationship between the mass of the black hole and the size of the event horizon, in the case of M87, we get a diameter of 38 billion km, less than that of our own solar system. At a distance of 55 million light years, this value corresponds to an angle of 15 micro-arcoseconds, roughly a 2 euro coin on the Moon. In fact, there is nothing to see at the event horizon; what matters is the "photon ring," the circle roughly twice the size of the event horizon generated by photons deflected by the gravitational field of the black hole that acts as a powerful gravitational lens. In addition, since the accretion disk rotates, it is reasonable to expect that the part coming towards us is strengthened by the Doppler effect and appears brighter.

Reaching such an extreme angular resolution seemed like a mission impossible, but astronomers do not balk in the face of anything and, thanks to a planetary effort, they succeeded in the enterprise. To make the most of the size of the Earth, the Einstein Horizon Telescope (EHT) succeeded in putting together 8 radio observatories scattered between the South Pole and Greenland, passing through Chile, Mexico, Arizona, Hawaii and Spain (Fig. 9.6). If everything worked out, they hoped to get to a resolution of 20 micro-arcoseconds, just above the expected size of the photon ring. In April 2017, the 8 telescopes observed M87 in perfect synchronization.

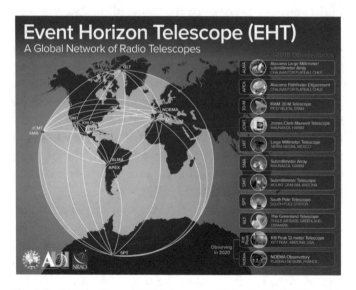

Fig. 9.6 Distribution of EHT telescopes Note that in 2018 and in 2019, few more locations were added to the original group of 8 telescopes (*Credit* EHT collaboration)

Then began the long work of correlating the data so as to arrive at an image that would not disappoint expectations. The first view of the monster's shadow in the center of the M87 galaxy was published in 2019, and was immediately dubbed the photo of the century. Thanks to an angular resolution that would allow you to see a credit card on the Moon, an asymmetrical donut appeared, very close to what was expected, but no less interesting for it. The emission comes from a region about four times the size of the orbit of Pluto that represents the outer boundary of the event horizon of a 6.5 billion solar mass black hole. The members of the EHT collaboration knew that their doughnut contained further information, directly related to the physics of emission, and set out to look for the presence of effects due to radiation polarization. It took them two years of data analysis to conclude that some of the detected radiation is polarized, that is, it has a preferred plane of vibration. This is the proof that the region where the brightest part of the radio flux originates must be pervaded by magnetic fields, indicated as streaks, which allow us to estimate the density and temperature of the plasma responsible for the emission. Astrophysicists who wondered whether the magnetic fields in the accretion disk around the black hole were chaotic or ordered now have an answer. The elegantly aligned streaks are evidence of non-negligible magnetic order in the accretion disk (Fig. 9.7).

Polarization is not only a prerogative of the areas surrounding the black hole. In the figure, we see, in succession, images with better and better resolution, starting from the Hubble Space Telescope up to EHT, passing through the polarized images obtained by the ALMA array and the VLBA combination (Fig. 9.8).

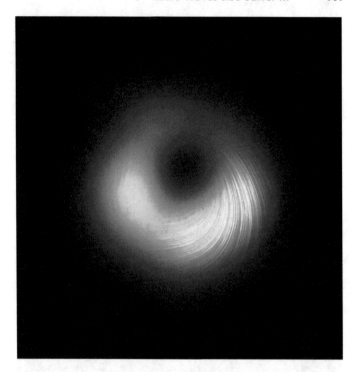

Fig. 9.7 A polarized view of the black hole accretion disk emission at the center of the M87 galaxy. The streaks indicate the direction of polarization, thus indirectly mapping the magnetic field that pervades the environment (*Credit* EHT Collaboration)

Deep Space Network

In addition to receiving the faintest of celestial signals, large antennas are used to keep in contact with the probes working in interplanetary space to study the Earth, the Moon, the Sun, Mercury, Mars, the asteroid belt, Jupiter, Saturn, Pluto and the periphery of the solar system. In addition to these, there are astronomical missions dedicated to the study of the universe through all types of

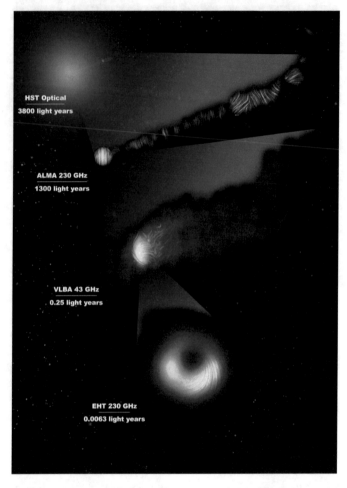

Fig. 9.8 An image of the M87 galaxy in visible light taken with the Hubble Space Telescope (top), followed by three images of the central region in polarized light. The three polarized light images zoom in closer and closer to the supermassive black hole; they were obtained with Alma (second from top), Very Long Baseline Array (Vlba; third from top) and Eht (last from bottom). The lines indicate the orientation of the polarization, which is related to the magnetic field in those regions. *Credits* Eht Collaboration; Alma (Eso/Naoj/Nrao), Goddi et al.; Nasa, Esa and the Hubble Heritage Team (Stsci/Aura); Vlba (Nrao), Kravchenko et al.; J. C. Algaba, I. Martí-Vidal

radiation that are absorbed by the atmosphere. But sending probes, with state of the art technologies, is not enough. Once placed in orbit, or on the right trajectory, each probe must be followed constantly so that it can receive orders and send the data it has collected back to earth. The exploration of the solar system, and of the whole Universe, depends critically on the ability to interact continuously with our robotic explorers. Losing contact with a probe means losing its full potential for discovery. Keeping tabs on the large family of interplanetary probes 24/7 is not easy. To enable continuous coverage of the sky, antennas must be carefully positioned on the ground. NASA, in collaboration with the European Space Agency (ESA), has chosen to place the ears of its Deep Space Network (DSN) at Goldstone (California), near Madrid (Spain) and near Canberra (Australia). Each DSN centre has antennas of different sizes: the largest ones are dedicated to listening to the most distant objects, the signals from which are weakened by the distance, while the others handle the nearest probes, with particular attention to Mars.

In addition, one has to consider that each probe has a very specific location and can only be seen at certain times from certain locations on Earth. Thus, each antenna's pointing sequence is prepared and continuously updated so that all active missions have their portion of listening time. To view the interplanetary communications traffic in real time, you only have to visit DSN right now (http://eyes.nasa.gov/dsn/dsn.html). As the hours (and days) go by, all the probes show up for their turn. It's a kind of interplanetary roll call, in which newly launched missions, like Perseverance, are antenna neighbours to old glories like the Voyagers, which, over 40 years after launch, are punctiliously listened to every day, because now that they've left the Sun's zone of influence, their data unveil the interstellar environment.

Obviously, each probe operates exploiting a different technology with different transmission capabilities. The data collected range from the very high-resolution images of Perseverance and Ingenuity (which required an upgrade of the Canberra antenna receivers) to the few bits per second of the historic Voyagers.

Since the antennas of the DSN are in every way similar to radio telescopes, radio telescopes equipped with the appropriate instrumentation can also be used to increase the ability to receive probe signals at particularly important, or simply crowded, times. In July 1969, for example, the radio telescope at Parkes, in Australia, played a key role in receiving the television signal of the landing of the Apollo 11 mission. With an eye towards a future that includes European explorations on the planet Mars, the Italian Space Agency equipped the Sardinia Radio Telescope with a receiver suitable for collecting data from the probes.

Radio Astronomy in a World Filled with Radio Waves

Radio astronomy is a very powerful   tool for the study of the cosmos, however, the fact that it is called *radio* suggests that we are talking about the same waves that are used to transmit the signals from radio, television, radar and mobile phones.   This immediately creates a serious problem of coexistence between radio astronomers and the communication society in which we all live, immersed in a dense network of radio waves that envelop the Earth. It is a pillar of our technological civilization, since everything we do depends on the possibility of transmitting information through radio waves. It's a pervasive but relatively recent phenomenon: in the early 1900s, there was absolutely nothing. Before Marconi found a way to use them, electromagnetic waves were little more than an equation, known to very few physicists. Now, they are an essential part of everyday life, and the demand for the use of radio frequencies is increasing all the time. But the available frequencies are of a finite number and cannot be used freely. There is an international body, called the International Telecommunication Union (ITU), which is responsible for allocating radio frequencies and which periodically meets to consider new requests.

Ambitious and very profitable projects, such as radio and television broadcasting and, more recently, mobile telephony, require dedicated frequencies that must not overlap with frequencies already allocated for other uses, such as radar or amateur radio, with particular attention being paid to institutional communications, for example, of the police, the military, rescue services or disaster relief. You will certainly have heard of frequency auctions, and you can imagine what interests are involved. Once a frequency band has been assigned, the ITU recommends that the signal emitted be precisely within that frequency range without overflowing to neighbouring frequencies assigned to other uses.

Since 1959, radio astronomy has been assigned precise frequency bands chosen so as to allow for the study of celestial bodies. The date is not accidental: in October 1957, the space age had officially begun with the launch of the Sputnik satellite, and, when the radio telescopes in operation at the time picked up the famous beep beep, it became immediately clear how powerful the signal was. However, I really don't think that radio astronomers were concerned about satellite transmissions; it was still a pioneering era. What worried them, rather, were the terrestrial transmissions that were growing at an impressive rate and risked cancelling out the astronomical signals.

The ITU decided to allocate 21 frequency bands for the exclusive use of radio astronomy. Obviously, the frequencies chosen were amongst those of greatest relevance for the understanding of the Universe. The most important of all is certainly the frequency at 1420 MHz, which corresponds to the line emitted by neutral hydrogen at a wavelength of 21 cm, as we have seen above. By measuring the radiation emitted at this frequency (and in its immediate vicinity), radio astronomers can map the structure of galaxies and understand how they move. More generally, radio waves allow us to reveal the presence of many molecules, each emitting a certain line at a certain frequency. Over 200 molecules are currently known, and more continue to be added. What does not change, however, is the framework of the frequencies assigned to radio astronomy, which are only 2% of the frequencies available to the terrestrial user, but must be kept free of human signals precisely to allow scientists to pick up the signals that come to us from celestial objects. The cleanliness of these bands is fundamental for radio astronomy research, which, let's not forget, involves a passive use of frequencies. Radio telescopes are often enormous structures, built to search for very weak signals. This capacity is, at the same time, their great strength and their intrinsic weakness.

The ability to easily detect the signal from a mobile phone on the Moon makes them vulnerable to the intrusions of any terrestrial signal that turns out to be millions (sometimes billions) of times more powerful than celestial radio sources. The fact that frequency bands are "reserved" for radio astronomy does not protect them from "losses" by transmitters working in adjacent bands and producing a signal that is a little less precise than it should be. For this reason, radio telescopes have an ad hoc professional figure (called a *spectrum manager*), who works with others users of the spectrum, devising preventive and corrective actions to limit electromagnetic pollution, and thus helping to safeguard radio astronomical research. In fact, it is impossible to work in the presence of noise a million times stronger than the signal one wishes to detect. We must also consider that celestial sources emit at a given frequency that cannot be shifted according to the needs of the observer.

I wrote earlier about hydrogen, the most abundant element in the Universe, which emits its line with a wavelength of 21 cm. This value was decided by Mother Nature, without the possibility of negotiation.

Thus, to map the hydrogen distribution in our Galaxy and elsewhere in the cosmos, the 1420 MHz part of the radio spectrum must be free of interference.

Radio Quiet Zones

Indeed, radio astronomers do not have an easy life. Although large radio telescopes are built in isolated places, mostly in large valleys, naturally protected by the surrounding mountains, there are many sources of noise. Geography is not enough to preserve the celestial signals. Around radio telescopes, *radio quiet zones* are established where it is not possible to use mobile phones, laptops, Bluetooth or WIFI connections, remote controls for garage openers, anti-collision radars, or microwave ovens (unless they are inside a Faraday cage that shields their radiation). These are all sources of interference that need to be eliminated. In the village of Green Bank near the Green Bank Radio Observatory, which is the largest steerable radio telescope in the world, people live in a spacetime fold where nobody can use cell phones or WIFI. A sign outside the building where the scientists work informs you that the radio telescope is capable of detecting radiation from your cell phone put in airplane mode all the way to Saturn, so no one is allowed to keep their cell phone on.

In Socorro, New Mexico, where an array of 27 radio telescopes, known as the Very Large Array (VLA), operates, cell phones left on by distracted visitors are located and owners are told to turn them off.

In China, a draconian solution was adopted. The 9000 people who lived within 5 km of the brand new Five Hundred Meter Aperture Spherical Telescope (FAST) were forced to move.

Radio Astronomers did not want to go to the effort of building the largest fixed telescope in the world (500 m in diameter) and then have to worry about interference with the residents' electronic equipment! We do not know what the reactions of the people who had to leave their homes were. However, it is clear that the lack of residents cancels the sources of noise: no people means no radio and TV signals, no cars, no mobile phones, no microwave ovens and no whatever else.

Radio Telescopes and Satellites

Institutional attention to radio frequencies and interference-free zones around the observatories are not, however, sufficient to guarantee the operation of radio telescopes. The most dangerous threat comes from above, where telecommunications, telephony and global positioning satellites orbit. Concerns about a difficult coexistence began in 1982, with the launch of the Soviet (later Russian) GLONASS global positioning satellites, which transmitted at a frequency of 1612 MHz, just next to the astronomical band of the hydroxyl radical (OH), one of the most common molecules in the Universe. As soon as one of the GLONASS satellites appeared on the horizon, the receivers were blinded. The problem was solved with scientific diplomacy involving spectrum managers and astronomers from all around the world convincing GLONASS not to use those frequencies, but it came up again in 1998, with the Iridium constellation, the first example of a satellite fleet dedicated to global telephony. The Iridium satellites were not transmitting in one of the radio astronomy bands, but were in the immediate vicinity, and their overseers had not adopted any system to be sure that the side lobes of their signal did not interfere with the hydroxyl radio band. And we are not referring to a small interference: Iridium was transmitting a signal 1000 times stronger than what the ITU had recommended.

To defend the receivers, an Iridium filter was developed that was meant to mitigate the disturbance via software, but the problem was never solved.

To further complicate the question, one should consider that satellites in low orbit are not the only source of radio interference. There are also those in geostationary orbit, which always appear stationary relative to the earth's observer. These are the satellites that distribute the signals for satellite TV channels, for example, and it is absolutely essential that we avoid pointing radio telescopes in their direction. It is a matter of studying the sequence of observations to avoid intercepting telecommunications satellites. This is a delicate task, but it can be done, because there are only a limited number of satellites and they stand firm in their geostationary orbit allocation. The coexistence of radio astronomy research and a hyper-connected world depends on the success of these efforts.

Unfortunately, however, the outlook is not good, because the situation is certain to get worse in the years to come.

The proliferation of new satellite constellations will create new challenges, and mitigation strategies will have to be devised on a case-by-case basis.

Given the commercial value of radio astronomical frequencies, it has been computed that selling radio-protected bands would provide a very interesting annuity to all radio astronomers.

Obviously, this is a paradox: radio astronomers do not own the frequencies, and therefore they cannot even imagine doing business with companies that are willing to pay handsomely for the precious allocation, without even remotely considering the hypothesis of investing in science, which, alas, does not yield immediate revenues.

10

An Increasingly Crowded Sky

For years, on summer evenings, my husband and our daughter's favourite game was to lie down on the terrace of our house (reasonably sheltered from the village lights in a quiet town in the Ligurian hinterland) and hunt for satellites. When night falls and it's already dark on Earth, the satellites, orbiting at a height of a few hundred kilometres, are still illuminated by the Sun, and the light they reflect makes them (more or less) easy to spot. Maybe they could also see some shooting stars, but that wasn't the subject of the competition between father and daughter, who challenged each other to see who could spot the most satellites, often managing to involve the neighbours. The terrace became an encampment, with large pillows, mattresses and blankets. Over the years, a scoring system had been refined in which the sighting of smaller, faster and less luminous satellites was associated with a higher number of points, compared to the slow, large and very luminous ones, because they were easier to see. Shooting stars

© The Author(s), under exclusive license to Springer Nature Switzerland AG 2021
P. Caraveo, *Saving the Starry Night*,
https://doi.org/10.1007/978-3-030-85064-7_10

were assigned the lowest score of all, because seeing those was just a matter of luck, not refined talent. In fact, the challenge was to see who could first detect the presence of a moving dot of light with brightness that changes, since the portion of a satellite's surface that is exposed to the Sun often varies, because they either rotate, as they were designed to do, or tumble uncontrollably. Once a satellite was spotted, there was the need to somehow indicate where the sighting had been made, because only a confirmed sighting would yield points. And then it was all, "It's there, there by that bright star on the right," "no, further down, see where that triangle of stars is?" and then, invariably, "there, see, it's passing through the Swan now," because it turns out that all satellites pass through the Swan, for the very good reason that they are, in general, Earth observation satellites that are in polar orbit and cross the sky in a North-South direction. I have never been good at the game. My eyesight isn't perfect, and my most important contribution was alerting the spotters to a possible International Space Station (ISS) passage. I was the first in my family to discover the power of apps that allow you to track the position of the most popular satellites, and I assure you that an ISS passage is always quite a sight. A football field covered with solar panels is sparkling in the sky. Moreover, thinking about the fact that there are astronauts on board makes you wonder. Obviously, if I had been at the console of a telescope making astronomical observations, instead of being on vacation, my attitude would have been radically different, and I would have used the same knowledge about the trajectory of the ISS to carefully avoid it. The passage of such a bright object would have ruined my observation.

In fact, optical astronomers have never had an idyllic relationship with satellites, mobile light sources that leave an imprint of their passage on the celestial image being acquired. Sometimes, the streak does not bother you too much, because the object you want to study is far enough not to be disturbed; sometimes you have to throw away the image and start over. However, it is a nuisance that does not occur too often and can be accepted among the contingencies of the profession.

These considerations were true until midnight on May 24, 2019, when Marco Langbroek, an avid satellite hunter, shot a video from Leiden, Netherlands, of a sequence of bright dots proceeding in a neatly lined up formation, almost looking like a little train (Fig. 10.1). They were the first 60 satellites of the Starlink constellation launched by SpaceX twenty hours earlier, cleverly stacked to form two towers in the fairing of a Falcon 9 rocket (Fig. 10.2). Each satellite weighs 260 kg and has a flat-panel chassis, about the size of a table (roughly 3 m by 1.5 m), where several antennas are housed, with a solar panel 9 m long, aptly folded at launch. Once they reach an altitude of about 300–400 km, the satellites are released in sequence and proceed neatly aligned.

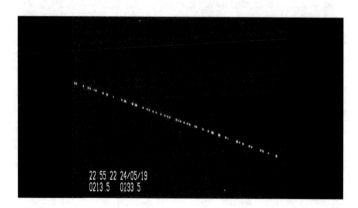

Fig. 10.1 Frame from the video by Marco Langbroek (https://www.youtube.com/watch?v=ytUygPqjXEc) that showed, for the first time, the luminous train formed by the 60 Starlink satellites launched a few hours earlier (*Credit* Marco Langbroek)

Fig. 10.2 A batch of Starlink satellites in launch configuration. They are stacked to form two towers, each consisting of 30 Starlink satellites (Credit SpaceX)

While it's dark on the ground, the structure of the satellites, as well as their big solar panels, which have opened, are still hit by the Sun and appear very bright in the sky as they pass one behind the other in a spectacular train-like sequence that many have perceived as a UFO. The much dimmer dots that, in the video, can be seen in the background are the stars that are visible to the naked eye, giving an idea as to how bright the Starlink satellites are. Roughly speaking, they are comparable to the Polar star.

The combination of the number, the relatively low orbit, the flattened shape (with the white, reflective antennas facing the ground), the inclination and the size of the solar panels has produced a deadly effect, capable of sending the astronomical observatories over which the merry little train has flown into a tailspin. The problem is particularly serious in the hours after sunset and before dawn, when it is dark on Earth but the objects in orbit are illuminated by the Sun. In the middle of the night, the problem does not arise, because objects in low orbit are in the Earth's cone of shadow and do not reflect.

The launch of the first group of satellites of the Starlink constellation was certainly not a surprise, to the extent that radio astronomers, more accustomed to having problems with satellite transmitters, had moved quickly to discuss their concerns with SpaceX, looking for possible ways to mitigate the satellites' effect. The optical astronomers, on the contrary, were simply unprepared for a problem of this magnitude. How could anyone have imagined that the Starlink satellites would be brighter than 99% of the satellites in orbit? It was only after seeing the bright "train" that astronomers began fully to appreciate the seriousness of the problem, knowing that it would only get worse as the number of satellites increased. Unfortunately, we're not talking about a possibility, but rather a sure thing. In fact, when Elon Musk says he wants to put thousands of satellites in orbit to provide a global internet service, he is terribly serious. After launching the first group of 60 satellites in May 2019, and then another in November 2019, he announced that, starting in 2020, SpaceX would launch a new batch of 60 satellites on a fortnightly or, at longest, monthly basis. The first group launched on January 6, 2020, the second on January 29, the third on February 17, the fourth on March 18, the fifth on April 22, and so on. In fact, since January 2020, the majority of SpaceX launches have been, and will continue to be, dedicated to the Starlink constellation. At the end of 2020, the minimum number of satellites was reached to be able to start testing the service with the help of a group of volunteers scattered across the U.S., Canada and, later, in England as well. This test is known as Beta Better Than Nothing, underlining the fact that it's a first attempt that offers a sub-optimal quality connection to a limited number of potential users of the service, which is aimed at the 12 million Americans (roughly between 3 and 4% of the population) who live in isolated areas that optical fiber still has yet to reach. In order to connect, users must use a special steerable antenna, and, in March 2020, the Federal Communications Commission approved SpaceX's request to operate up to 1 million antennas in U.S. territory for 15 years. The antenna is sold, ready to use, for $499. Is this the beginning of satellite internet? We'll see…

I confess that, even though I have a heart that beats for astronomy, I can't but admire the efficiency of SpaceX, which has managed to maintain an impressive launch rhythm for its Falcon 9, using the same ramps that were built for the Apollo program. Having spent almost my entire career following the design, realization and management of satellites dedicated to the study of X and gamma astronomy, I am very familiar with the problems, the difficulties and the risks of space research. I'm a fan of satellites, with which I have obtained my most important results, and I really appreciate SpaceX's original and innovative approach, which has changed the space business in a very short time.

If you read the launch reports, you will always find mention of the number of times the first stage, the one that is recovered through the very elegant controlled landing manoeuvre, has been used. With more than 100 Falcon 9 s having been successfully launched, the news is no longer the recovery of the first stage, but rather the few cases when the landing fails. Consulting the news, I see that, in the launch of March 14, 2021, the first stage was at its 9th ride. In addition, the fairing is also recovered and reused.

These are all pieces of a brilliant strategy, since the success of such a massive undertaking depends on the economy of scale, as well as the actual number of users. Indeed, Musk's first goal is to avoid going bankrupt, as happened to competitor OneWeb, which found itself without liquidity just after the launch of its second set of 30 satellites on March 18, 2020. This launch was only the third of 22 agreed upon in the contract with Arianespace to take the 650 satellites originally planned for the constellation into orbit at an altitude of 1,200 km, but the machine jammed. It appears to be a replay of what we might call the curse of the constellations. It is a very contagious syndrome that, in the last 20 years, has affected Iridium, Globalstar, Orbcomm and Teledesic, which have all gone through a financial crisis with bankruptcy proceedings. Some have since recovered, others have not. So, it's understandable that Elon Musk's first goal is to avoid the contagion spreading to his Starlink, which requires large investments, but still has yet to yield revenues. Luckily for him, however, SpaceX, which has to produce six Starlinks a day to keep up the pace, is supported by the burgeoning launch market (as well as government incentives to help connect areas that are not served by optical fiber within the United States), and this allows it to absorb, at least in part, the huge costs of building the orbital infrastructure. Unlike SpaceX, OneWeb is financed by shareholders all over the world. While headquartered in England, directors are distributed among the United States, Germany, Israel and Mexico. However, with 74 satellites in orbit, and many others already built, OneWeb, after declaring bankruptcy, went on the market and was bought for a billion dollars by a group of investors led by the British government and the Indian telecommunications company Bharti Global, who were determined to complete the constellation that, let's not forget, been assigned a radio frequency band for its operations. In fact, the new consortium, whose shareholders include Qualcomm (Singapore), SoftBank Group (Japan) and 1110 Ventures (U.S.), promptly asked for permission to launch more than 48,000 satellites, a piece of news that has understandably created concern in the astronomical community towards which OneWeb had always been rather deaf. The figure was later scaled back to 6,372 satellites, but the concern did not fade away.

In July 2020, Amazon was given permission to launch 3,236 satellites to provide satellite internet anywhere in the world. Their constellation is called Kuiper.

Combining the Starlink and OneWeb numbers with the small Iridium and Globalstar constellations, which are already operational, and also considering the plans of Facebook, Link, and the Canadian companies Kepler and Telsat, as well as companies in Russia and China, one could end up with 100,000 satellites in low orbit. That's a scary number.

Business Plan

Certainly, the idea of providing satellite internet at a reasonable cost to all parts of the world that aren't reached by land-based services is visionary, like many of Musk's ideas, and, no doubt, he has done his homework. The sale of internet services is a very fast-growing industry: it made $687 billion in 2019, and is projected to reach $74 trillion by 2026. In other words, it's expected to expand by more than a factor of 100 in 7–8 years, and the growth will come, in part, from new connections by potential users living in rural areas with little or no coverage. We're talking about two billion people spread across different continents. To get an idea of the figures at stake, Morgan Stanley has estimated that the turnover of the space industry, which currently touches 350 billion dollars, will be 1.1 trillion by 2040, 410 billion of which will come from the supply of internet services.

This is a staggering business that has attracted the attention of many other operators planning their own satellite constellations, some more numerous, some less. The initial investments are certainly important, but the expected revenues should amply repay them, since, in Musk's vision, they will be used to finance human exploration of Mars, his ultimate goal.

How many users might reasonably be interested in subscribing to this service? A recent study conducted in an academic environment, and thus completely independent from the companies involved, used 2018 data provided by UN and the World Bank to guess the number of potential subscribers. The study started by ranking the countries of the world according to the percentage of the resident population that has access to the internet, recognizing a need for satellite coverage in countries where less than 75% of the population is online. The second step was to estimate how many potential users could afford the cost of satellite connectivity, using an indicative value of $60 per month, and requiring that the annual cost of the service be less than 10% of the gross domestic product per capita of that particular country. Obviously, these are working hypotheses, perhaps even optimistic ones, since the real cost of the subscription to the Starlink service seems to be around $99 per month, and they are most certainly simplified, since they do not include the *convenience or curiosity* factor that could push a user, potentially reachable by another internet service, to prefer the satellite option. In any case, in a world unfortunately divided between rich and poor countries, it is clear that the latter are the most interested in the service, which, however, is too expensive to become effectively accessible. Leaving aside the "rich" countries, such as most of Europe, the United States and South Korea, which already enjoy a wide variety of Internet access (but might subscribe for other reasons), the service could be of interest to Turkey, Brazil, Mexico and China. In fact, however, the potential Chinese market is unlikely to turn to SpaceX. Those familiar with the topic say that a Chinese network of 13,000 satellites is under study.

The results of the study would not seem very encouraging for SpaceX and all other satellite networks. However, the situation could evolve to meet new needs, such as self-driving vehicles or IOT (the Internet Of Things). Certainly, SpaceX is not only looking at residential subscriptions and is trying to expand the number of its potential customers by offering an Internet signal on the move to meet the demand for connectivity that comes from the market of cars, trucks, ships and planes. To this end, SpaceX has applied for a license to operate what it calls *Earth Stations in motion* suitable for being mounted on planes, ships, trucks and, in the future, even cars. This would allow passengers to be connected while traveling, increasing the number of potential users of the service.

Starlink is also building a partnership with Google Cloud that will exploit the satellites' high-speed broadband internet connectivity to deliver data, cloud services and applications to potential customers in locations at the network's edge. Starlink ground stations will be located within Google data centres to take advantage of Google Cloud's high-capacity private network, enabling the delivery of critical enterprise applications to virtually anywhere in the world.

In parallel, the Starlink network could be of interest to the U.S. Department of Defence, which is studying the possible use of LEO constellations to strengthen its Global Positioning System (GPS), which is a military infrastructure based on a network of 32 satellites orbiting at an altitude of 20,000 km.

Each satellite continuously transmits its position and an ultra-precise time-stamp. Receivers on the ground compare the arrival and departure times received from all visible GPS satellites and compute their positions with an accuracy of a few meters. The GPS signal, which is vital to both our security and our economy, arrives on the ground attenuated by the height of the orbit, making it a possible target of jamming by hostile powers. Aware of this problem, the Pentagon has sought alternative solutions by looking at other satellite networks that could provide a service independent of the GPS network. First, they tried to see if the existing Iridium network could be considered, after which the inception of the Starlink constellation opened up new horizons, since a GPS device mounted on Starlink satellites could provide a positioning that would be both more accurate and based on a much stronger signal, and therefore almost impossible to jam.

In fact, the signal produced by a low-orbit satellite would be about 1,000 times more intense than that of GPS satellites, which are much further away. Moreover, a network based on tens of thousands of satellites, as Starlink's should be, would be much more difficult to knock out, because it can rely on considerable redundancy. In addition, using the latest generation of GPS devices, coupled with the computing power available today, it would be possible to achieve localization accuracy on the order of centimeters, rather than meters. Adding the GPS device on board would require less than 1% of the transmission capacity of the Starlink satellites and would consume 0.5% of the available power. However, having such precise positioning would not come without a price for the user, who would need a receiving station for the Starlink signal, as well as upgraded instrumentation. This is a consideration that does not frighten the military, who have invested at least 12 billion dollars in their GPS network, the management of which costs about 2 million dollars per day.

If the Pentagon wanted a robust alternative to GPS, one could say that Starlink is an interesting possibility. Elon

Musk won't let the opportunity slip away, just as he took the chance offered by the U.S. government that granted him generous subsidies to increase the connectivity of rural regions.

Satellite Constellations and Astronomy

To be a pioneer implies both honours and burdens. For this reason, without wanting to point the finger at SpaceX, I will use Starlink as a case study to describe the problems created for astronomic research by the satellite constellations, with the proviso that other constellations will certainly come into existence and each one will have its own peculiarities.

First of all, let's remember that we are talking about LEO constellations, that is, the ones formed by satellites in Low Earth Orbit. They represent a choice completely different from that adopted by those who use the classic telecommunications satellites that operate in much higher orbits, with an ideal positioning within the geostationary orbit. Satellites in "high" orbits have the advantage of seeing a large part of the Earth, and therefore a limited number of satellites are sufficient to cover the entire globe, as in the case of the previously mentioned GPS constellation, composed of 32 satellites (24 of which are always active) at an altitude of 20,000 km. However, the height of the orbit implies non-negligible waiting times due to the transit of the signals that have to go from the user to the satellite and then back from the satellite to the user. In addition, the signal produced by distant satellites is weakened by their distance. Considering the pros and cons of the different orbits, SpaceX opted for "low" orbits, at a height of about 550 km, that allow for the provision of fast service without appreciable latency times for the user, but that require the use of a large number of satellites, the reflections of which are definitely bright (because the brightness of a source is proportional to the inverse of the distance squared).

The launch of the first set of Starlink satellites raised red flags for optical astronomers. After devoting so much effort to building telescopes in the best places in the world, though certainly not the most convenient to reach, astronomers could hardly have imagined that danger would come from above.

After the protests following the May 2019 launch, Elon Musk first tried to downplay the issue by pointing out that, with thousands of satellites already in orbit, his first batch of 60 Starlinks couldn't be causing all that disruption. Too bad for him that the visibility (or, if you prefer, the disturbance) of a satellite depends a lot on the height of the orbit, as well as its shape and the material used to build it. Starlink satellites are in low orbit and resemble large tables. It's this strange flattened shape, designed to maximize the number of satellites contained in the launcher's fairing, that makes them so reflective. Certainly, among the conditions considered by the designers, minimizing the reflectivity of the satellites was not something that anyone had asked about, a problem that only promises to get worse, as their number is constantly growing. With the February 2020 launch, which brought the Starlink satellite count to 300, SpaceX had already become the majority shareholder within the "low" orbits, having a fleet larger than the sum of all the other active satellites orbiting less than 600 km from Earth, with a presence that grows by roughly 100 units every month.

Commenting on Marco Langbroek's video (and all those that followed it), Musk pointed out that the light train effect is momentary. Just after launch, when the satellites are released at an altitude of about 350 km, the large solar panels are deployed in the "open book" configuration parallel to the ground, so their reflected light adds to that of the satellite's body. Then, the orbit is raised to 550 km, thanks to an ion engine: the distancing decreases the brightness, which further decreases with the change in orientation of the solar panels, which transition from the open book configuration (parallel to the ground) to the shark fin configuration (perpendicular to the previous one) (Figs. 10.3 and 10.4).

In this way, the solar panel would be hidden behind the body of the satellite, which would become the only source of reflected light, too dim to be visible to the naked eye. In addition, the repositioning operation disperses the satellites so that they no longer appear all together. However, even if they won't be visible to the naked eye, they will continue to appear in astronomical images, where their presence will be a constant in the hours just after sunset and before sunrise.

In fact, SpaceX didn't shy away from the confrontation with astronomers, and early January 2020 saw the launch into orbit of DarkSat, a test satellite, painted black to be less reflective. Measurements made in March, when the blackened satellite had reached its final orbit,

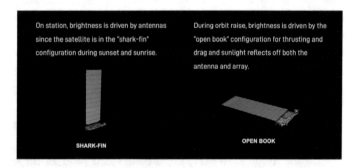

On station, brightness is driven by antennas since the satellite is in the "shark-fin" configuration during sunset and sunrise.

During orbit raise, brightness is driven by the "open book" configuration for thrusting and drag and sunlight reflects off both the antenna and array.

SHARK-FIN

OPEN BOOK

Fig. 10.3 Schematic of the two different solar panel configurations of the Starlink satellites. On the right is the "open book" configuration used after launch and during the orbit-raising manoeuvre. On the left is the shark fin configuration used when the satellites are operational (*Credit* SpaceX)

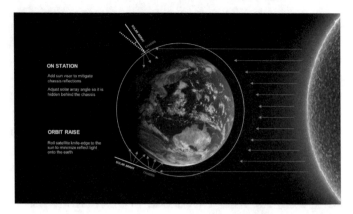

Fig. 10.4 Schematic view of the different geometry of the reflecting surfaces of the Starlink satellites, at the critical moments of sunrise and sunset, just after launch (below) and during operations (above) (*Credit* SpaceX)

showed that its reflectivity had been halved. This is good news, but not enough to allay the concerns of astronomers who are convinced that more effort must be put forth to make the satellites even less reflective. In addition, blackening, though promising, may not be a perfect solution. Satellites always have problems of thermal control and have to disperse the heat produced by the on-board electronics. By limiting reflectivity, blackening makes them more absorbent of solar radiation, which could overheat them. As if that weren't enough, a hot satellite becomes brighter at infrared wavelengths, a part of the spectrum that also interests astronomers. In fact, as an alternative to DarkSat, SpaceX has developed a system of sun flaps designed to prevent the Sun at sunset, i.e., at near-grazing incidence, from reflecting off the surface of the satellite. VisorSat, the satellite with two sun visors designed to shade its antennas, seems a simple and effective solution for lowering the reflectivity of Starlink satellites while not interfering with their structure, given that they are made in series and must be able to be stacked in the launcher fairing. Indeed, believing in this solution, SpaceX decided to send only satellites equipped with sun visors into orbit, starting from the launch of August 2020. SpaceX's decision was based on simulations, prior to the verification of the effectiveness of the solution, which relied on data collected between August and December 2020, released in January 2021. Based on more than 1000 measurements, we now know that the average magnitude of a VisorSat is 5.93 compared with the value of 4.63 measured for the other Starlinks, with a reduction of 1.3 magnitudes. In other words, the brightness of a VisorSat is about 30% of that of satellites without sun visors. The reduction was also achieved by optimizing the orientation of the solar panels to try to minimize reflection. This means that, while the original satellites could be seen with the naked eye when they reached their final orbit at 550 km altitude, the VisorSats are too faint to be seen by naked-eye observers. This too is good news, though astronomers hope that more can be done towards the goal of lowering the reflected light even further so as to reach magnitude 7. This request was prompted by the consideration that the measurements refer to satellites in their final orbit, while, given the number of launches that must continue for years, there will always be hundreds of satellites in lower orbits, either because they're going up to start their work or because they're coming down at the end of their mission and are preparing to re-enter the atmosphere, where they'll be destroyed.

I'd like to point out that the Starlink satellites are just the first on the list, and while other constellations are getting off the ground, launches have already begun for the OneWeb constellation, the managers of which have never shown much sensitivity to the astronomical world.

Even if no one has *certified* any solutions, it is important for everyone to understand that we are just at the beginning: satellite internet is very attractive and, if it works, it will become competitive with terrestrial services. It is no coincidence that there is a queue to build constellations in orbit, although the financial challenge of this type of operation is far from trivial. The case of OneWeb, as well as those of all previous constellations, does not discourage giants such as Amazon, Facebook, and Samsung, just to name the most well-known ones, who have equally ambitious plans for a total of tens of thousands of satellites in orbit at a height of between 500 and 2000 km. Additionally, the telecommunications companies that don't have plans involving constellations are betting on those of others: Vodafone, for example, has already invested in SpaceMobile with the idea of ensuring connection everywhere.

How Bad is the Problem?

As was done in the case of light pollution, the first action to take is to get a clear idea of the extent of the problem, going beyond the visceral effects of light trains, which are so spectacular that every Starlink launch causes a spike in UFO sighting reports.

Again, the approach is twofold, with volunteers on one side and professional astronomers on the other.

Simulations of the planned Starlink constellation show that, especially in the belt near the horizon, the perpetually moving points of light produced by Starlink satellites would be far more numerous than the stars that are visible to the naked eye, altering the view of the sky, which, let's not forget, is part of the heritage of all humankind.

With the precise goal of objectively estimating the deterioration in the quality of the sky visible to the naked eye, NASA has launched a Citizen Science project that asks for photographs of the Starlink satellites, both as they form trains and as they rise and move apart from each other.

Photographing Satellite Crawls

The Satellite Streak Watcher project aims to document the growth in the number of satellites by asking participants to use their cell phones to photograph the way in which the streaks produced by the passage of satellites increase over time. The satellites move at a speed of 1° per second (about twice the diameter of the Moon every second) and are bright enough to be photographed with a mobile phone at the beginning and at the end of the night.

It is advisable to attach the phone to a stand, choose an exposure time of 10 s (corresponding to a 10° strip in the sky) and look for the best setting for night photos.

Once everything is ready, you can go to *Heavens-Above. com (or use the appropriate APP on your phone)* and pick Starlink from the list of satellites to get the list of passes, complete with the transit time and expected magnitude. Select the brightest (the one with the lowest mag), take pictures during the whole transit, and then choose the best ones to send. This exercise should be repeated many times over a period of months so as to follow the evolution of the situation as time goes by.

In parallel, astronomers have been working to assess the impact of satellite constellations on the quality of the data collected by their telescopes. It must be said that telescopes are not all the same; besides the size of their primary mirror, the most important parameter in this study is the size of the field of view. Both the telescopes that have "small" fields of view and those with "large" fields of view were considered. As it was reasonable to expect, small fields of view are less affected by the presence of satellites, because they can avoid observing that area near the horizon where, just after sunset and before sunrise, they are more likely to run into satellite crawls. Fortunately, the large European telescopes in Chile shouldn't be too affected, also because astronomers, in general, tend to avoid observing near the horizon, since the quality of the images is never perfect, owing to the atmospheric turbulence. The situation is completely different for large field telescopes, such as the Vera Rubin Observatory, also in Chile, which was precisely built to observe large regions of the sky. Each image will cover an area corresponding to 40 times the full moon, and the telescope will struggle to avoid the satellites' crawls. Unfortunately, about 1/3 of the images will be crossed by a satellite that will leave a streak that is very difficult to eliminate with cleaning algorithms, because it's too bright. It will be necessary to delete the streak, losing all of the information contained behind it, hoping that the electronics, disturbed by the intense signal, will not produce other ghost streaks that would obviously worsen the situation. It is not clear how much of a "ruined" image can be saved. Certainly, the damage will vary from case to case, but it will never be zero and, besides the loss of information, one has to take into account the time needed to develop and then apply corrective software. Considering that the Vera Rubin aims to do large celestial surveys, that is, very extensive and very uniform coverage of the sky, losing one image out of three is difficult to accept. It is not worth building an extraordinarily sensitive and innovative instrument just to have it blinded by a new technology. A reasonable compromise must be found, and astronomers are willing to give it a try.

An attempt was made to evaluate the possibility of programming the telescope's pointings sequence in order to avoid the satellites. In parallel, astronomers tried to understand if it is possible to close the shutter of the instrument (suspending data collection) during satellite passages, especially when they are in train configuration. In this way, crawls could be avoided, provided some loss of observation time is accepted. All of these actions imply precise knowledge of the orbit of each satellite. Thus, astronomers have decided to build SatHub a repository to collect all the relevant (and available) pieces on informaton on the growing satellite population. It is certainly a wise decision, but the ability to take pre-emptive action so as to avoid the worst is limited. Unfortunately, there are so many satellites planned for the next few years that the task seems almost impossible. The only solution is to limit the amount of reflected light and, as we have seen, this is the strategy that is being pursued.

However, the astronomers' simulations led to an important result, showing, in a very convincing way, that, in addition to the physical characteristics of the satellite, what matters is the height of the orbit, which should never exceed 600 km. Satellites orbiting at a higher height, although farther away, and therefore intrinsically weaker, move more slowly, and thus their crawls are no less evident than those of satellites in low orbit, but they have the unpleasant characteristic of remaining visible all night long, since they never enter the Earth's cone of shadow.

Astronomers are therefore very clear in their request that the operators of orbital constellations, besides working to minimize the reflectivity of satellites, only use orbits with an altitude of less than 600 km. If we consider that OneWeb wants to operate its thousands of satellites at an altitude of 1,200 km, it becomes clear how really very worrying the situation is, especially because we cannot accept the future of world astronomical science depending on the goodwill of investment groups that are much more sensitive to economic returns than to the preservation of the sky.

Satellite Photo Bombing

I won't deny that light trains can be fun to watch. With the help of the information at the Heavens above website (and APPs), you can know in advance when the satellite will be visible at your location, and thus you'll have the chance to prepare yourself while avoiding the need to worry about a possible UFO visit. Do not rush, make yourself comfortable, give your eyes the time to get accostumed to the dark and look up at the sky, so you can appreciate how much the satellites' train can alter the sky's appearance.

If you want an example of the effect of the passage of the Starlink satellite train, look at the photo below, taken from a field of sunflowers in Brazil (you can tell that you are in the southern hemisphere because you can see the Magellanic Clouds, two small satellite galaxies of the Milky Way). The sky, which appears to be of pretty good quality, and has been spared the pollution of artificial lighting, is nonetheless marred by the cluttered presence of the bright satellites. In the upper right corner, there is also a very bright shooting star (Fig. 10.5).

At this point, imagine what could happen with thousands of satellites in orbit, when, in the hours after sunset and before sunrise, hundreds of satellites will be visible low on the horizon, turning the sky into a swarm of lights. Moreover, to provide the fast service they promise, internet satellites must fly over every part of the planet, even the most remote and inaccessible areas, where astronomers have taken refuge to escape light pollution. When this happens, the result is dramatic, with images (like the one in the picture) crisscrossed by dozens of streaks too bright to be corrected by clever software (Fig. 10.6).

And the situation can only get worse.

Fig. 10.5 Photo taken from NASA's Astronomy picture of the day (December 10, 2019) site. The author of this image (which results from the sum of 33 exposures) is Egon Filter (*Credit* Egon Filter)

Even Small Debris Are Harmful

So far, we have focused on light trails produced by the reflection of the Sun's rays on the body of a satellite (or on its solar panels). Let's now try to consider the cumulative effect of everything orbiting around the Earth, including both "big" and "small" objects, since, besides direct reflection, there is also the phenomenon of sunlight scattering. We have already touched on the scattering of light by the aerosol in the atmosphere when we mentioned that it is responsible for the formation of light halos due to the ill-directed and excessive artificial illumination. Now, let us examine what happens when a large number of orbiting objects, many of which are classified as space debris, are hit by sunlight's photons that subsequently scatter it in all directions. The portion of radiation directed downwards enters the atmosphere and is diffused by molecules

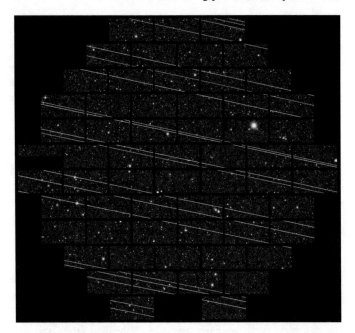

Fig. 10.6 Large field of view Astronomical image ruined by passing Starlink satellites (*Credit* CTIO/NOIRLab https://www.nsf.gov/NSF/AURA/DECam DELVE Survey)

and dust, thus increasing the background brightness of the sky. A group of researchers tried to model all of the objects surveyed in orbit (neglecting those that were too small to be measured and counted) to assess their contribution to the sky's brightness. Without considering the new constellations, it was found that the contribution of everything that is already in orbit to the background brightness of the sky at zenith is about 10% of the natural value. This is not good news: the value is far from negligible, especially if we consider that no part of the world can be protected from this pervasive pollution, which is, unfortunately, bound to increase with the new satellite constellations. Having built an observatory in a remote location does not help, because we are all under the same sky. In fact, in 1979, the IAU picked the value of 10% as the limit of the contribution of diffuse light, compared to the background brightness of the sky, when determining the suitability of a site for the construction of an astronomical observatory. If the authors of the study are right, there really is cause for concern, especially since the light scattered by orbiting objects will increase in proportion to their number, a consideration that makes the effort to reduce the brightness of everything we put into orbit all the more important.

The Most Penalized Science

While astronomers are right to fear that satellite intrusions will penalize every branch of astronomical research, it is clear to all that the most affected field will be the search for potentially dangerous asteroids, because they describe orbits that may bring them too close to Earth. The solar system hosts billions of asteroids, from the largest, like Vesta and Ceres (which have been classified as dwarf planets), to objects (more or less compact) of a few metres. The vast majority of asteroids orbit between Mars and Jupiter in what is called the asteroid belt. They are interesting from an astrophysical point of view, because they preserve information about the birth and evolution of the solar system, but they are not the dangerous members of the large family. The potential dangers come from asteroids moving in orbits intersecting that of the Earth, known as NEOs (for Near Earth Objects).

The basic idea is very simple: given two celestial bodies moving along different orbits that happen to cross each other, it is possible for the two objects to be in the same place at exactly the same time, and thus collide. The Earth is continually being hit by celestial rocks, adding up to as many as 40,000 tons per year, but they are, for the most part, small objects that burn up in the atmosphere. Every now and then, something hits the ground, and these are what we call meteorites; their size varies from pebbles to boulders. Meteorites are reported to have pierced the roofs of buildings, even hitting people, but no one has ever died from a celestial rock. This also holds true for the spectacular Chelyabinsk event, in which, on February 15, 2013, a celestial rock exploded with a great roar on a cold Siberian morning. Nobody was hit by any meteorite fragments. In fact, all resulting injuries were attributable to glass broken by the shock wave from the explosion. This was an experience that everyone involved would have gladly avoided and one that they will not remember with pleasure.

To get an idea of the number of visits by celestial stones recorded over a period of 10 years (from 1994 to 2013), let's look at this planisphere, which summarizes all of the events recorded in the atmosphere. The size of the circle is proportional to the intensity of the event that produces a flash of light as the alien body burns in the atmosphere. The events recorded at night are in blue, the daytime ones in yellow. The Chelyabinsk event (due to a stone about 15 m in diameter) certainly stands out as the largest in regard to size, but there are many others that are slightly less intense, mostly occurring over the sea, and revealed by the automatic network that measures the sonic shock waves that are generated (Fig. 10.7).

Counting celestial fireballs gives us an idea of the number of hits that we get, but it doesn't tell us if there are asteroids on a collision course with Earth. Because it is unpleasant to be under an unknown threat, networks of automated (and non-automated) observatories have been set up to take a census of NEOs and identify asteroids with orbits that are dangerously close to Earth's. NASA (along with other organizations) has invested heavily in

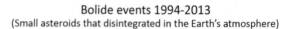

Bolide events 1994-2013
(Small asteroids that disintegrated in the Earth's atmosphere)

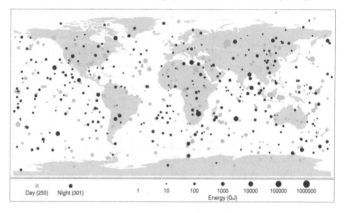

Fig. 10.7 Census of meteorites that were seen flaming in the atmosphere between 1994 and 2013. Yellow circles refer to events that occurred during the day, blue circles to events that occurred at night. The size of the circle is proportional to the event's brightness (*Credit* NASA/Planetary Science)

this research. In fact, while it is easy to follow large asteroids (like the one responsible for the extinction of the dinosaurs 60 million years ago, thought to be 1 km in diameter), in order to see and classify the smaller ones, telescopes and dedicated programs are needed. While, at the beginning of the millennium, the catalogued asteroids were fewer than 1,000, today, we have exceeded 25,000, and as many as 3,000 were added in 2020 alone, proving that, despite the forced closures due to the pandemic, the techniques that have been developed are effective.

None of the catalogued asteroids is on a collision course with Earth, but that's not enough to completely reassure us. We know that the asteroids that have escaped the census could be potentially dangerous. That's why the NEO mapping programme continues, looking for them where they are most likely to be in the plane of the ecliptic where all the planets orbit. This means scanning the sky toward the horizon, just after sunset and before sunrise, with wide-field telescopes so as to catch fast-moving sources. Unfortunately, that is precisely the direction in which, and time when, the bright satellites are most annoying.

Astronomers are also concerned about the impact of satellites' streaks on the discovery space of astronomy. Quantifying a discovery potential is very difficult, but astronomers know very well that a gamma-ray burst, a gravitational wave, or any other very brief phenomenon must be followed in real time. Telescopes are designed with the ability to be repositioned very quickly, precisely so as not to miss a single moment of something that is certainly unrepeatable. No one wants to have to reckon with a satellite mark superimposed upon what could have been a discovery. The science of the unexpected (serendipitous) is too important to risk falling into jeopardy.

Radio Astronomers also Fear Constellations

So far, we have discussed the problems that plague all those who "look" at the sky resulting from the reflections of satellites in orbit; now, we move on to consider aspects more intimately related to the operation of satellites that, to provide global connectivity, must operate in the microwave band, transmitting powerful signals at frequencies close to those protected for radio astronomy.

The issue that radio astronomers have with the new constellations is twofold. First, they worry about the 10.6–10.7 GHz protected band; second, they are concerned about the possibility to use for their observations wider parts of the non-protected radio spectrum.

However, since the Starlink transmission frequencies have been used by satellites for a long time now, it is important to understand why the interference from a constellation of satellites could hurt radio astronomy.

When a radio telescope observes a celestial source, it sees a very small piece of the surrounding sky through its main beam, which is normally quite narrow, and the rest of it through its side lobes. Therefore the power of a constellation of satellites to interfere with a radio telescope comes from the instantaneous collection of the power received from all satellites above the horizon as detected by the radio antenna, with a very large gain in its main beam and a very small gain in its side lobes. Due to the extreme gain needed to detect a radio emission of celestial origin, a satellite in the main beam will surely blind the receiver, and one seen through its side lobes will be detected as a strong source.

Previously, all of these frequencies were normally used by a limited number of geostationary satellites, the locations of which are fixed in the local sky, making it possible for radio observatories to avoid pointing in their direction while also successfully collecting data in the non-protected spectral band from sources far from the geostationary satellite positions.

Now, this situation has drastically changed. LEO satellites move at great speed compared to the rotation of the Earth, and they can be found in any position in the local sky (depending on the constellations' orbital parameters and the latitude of the radio telescope). Observatories with receivers sensitive to 10.7–12.7 GHz are likely to see an increase in the noise baseline, and will definitely see repeated strong blinding signals when a direct encounter with a satellite happens.

The *spectrum managers* of the radio astronomical observatories are very busy protecting the frequency bands allocated to radio astronomy by the ITU. At the same time, they are also worried about the possibility of listening to celestial radio signals using non-protected frequencies and trying to negotiate with each constellation, since each one uses a certain transmission technology.

And they don't always find ears willing to listen. It comes down to scientific need versus the commercial need of an aggressively profit-seeking market.

Both Starlink and OneWeb use, for their downlinks, the 10.7–12.7 GHz frequency band, which is contiguous to the protected 10.6–10.7 GHz radio band intended for continuum observations by radio astronomy and other passive users of the spectrum. Since the signals produced by satellites are billions of times more intense than those we receive from the most distant celestial bodies, even if a small percentage of the satellite signal gets spilled over the radio astronomy band, it would be devastating for radio astronomy. Hence, the efforts to alert the regulatory entities and the constellation operators to the need to make sure that their signal is clean and does not trespass.

The one that should worry us the most is Starlink, both because they were the first and because theirs is the most capillary constellation, which, to ensure universal connectivity, must always guarantee that its satellites will fly, and transmit, everywhere, even over the most radio-quiet places in the world, where large radio astronomical observatories operate (or will operate). SpaceX, however, immediately proved to be sensitive to the requests of radio astronomers, who are seeing very positive signs coming from the operator. Establishing a contact with OneWeb was more difficult, and the representative of the NRAO (National Radio Astronomy Observatory) had to raise his voice, reminding the company that its constellation would not be allowed to begin operating in the absence of compliance with the obligations to minimize the interference with radio astronomy. His argument seemed to have been convincing and contacts were begun, after which, unfortunately, OneWeb changed ownership and multiplied the number of satellites planned, complicating the management of the problem.

The coexistence between very sensitive passive instruments and very powerful mobile transmitters becomes even more difficult when leaving the frequency bands reserved for the exclusive use of radio astronomy to enter those bands allocated for shared use. It is clear that, in these cases, the efforts to reach compromise solutions are even more important, because the ITU recommends that the services sharing the same frequency bands not interfere with each other, and finding satisfactory solutions depends on the goodwill of all.

Remember that radio astronomy is in an intrinsically weak position. A radio telescope can only receive, so it cannot cause any disturbance to satellite services, while, on the other hand, any emission from a transmitter in orbit can be a potential source of electromagnetic pollution.

Of course, the problem of satellite interference can be mitigated in several ways, and different solutions should be considered. Iridium Next, for example, has said that it is prepared to switch off its satellites when they fly over the field of view of radio telescopes. It is necessary to announce the sequence of pointings foreseen in advance so that satellite management can calculate when one of them enters a telescope's field of view and turn it off. Easier said than done, both because radio antennas have maximum sensitivity in the pointing direction and because they also "see" through their side lobes. So, annoyance can also arise from satellites passing out of the main field of view but entering the antenna's side lobes.

Thus, it is clear that the concept of a radio quiet zone must be extended to the space above the radio telescopes;

only in this way will it be possible to preserve the scientific potential of the instruments already operating and of those, much more ambitious, that are under construction. Let us think of the Square Kilometer Array (SKA), which will operate 197 radio telescopes in the Karoo Desert in South Africa and 131,000 antennas in the Australian bush. What would be the point of building the most sensitive radio observatory ever designed, in the most radio-quiet places on the planet, only to have observations continually disturbed and/or interrupted by the passage of swarms of satellites?

To make the situation even more complicated, it must be added that radio astronomers are watching, with great concern, the growing interest in the development of satellite constellations that exploit SAR (Synthetic Aperture Radar) technology to obtain hourly radar maps of the Earth. Although radar images are less sharply defined than those obtained in the visible, they do offer some advantages, because radar signal propagates the same way in both day and night and is not inhibited by clouds. Again, this is the commercialisation of a technology that, until now, was only used by a limited number of satellites operated by space agencies. The transmitters used by SAR constellation satellites produce a signal 50 times stronger than the maximum level manageable by the receiver of a radio telescope.

In other words, if the receiver is illuminated by such a powerful signal, it will burn out. Hence the requirement for orbital radar operators to avoid pointing their instruments towards regions around radio instruments. If a radar image of a radio astronomical area is needed, telescope operators should be notified in advance so that they can turn off their instruments to prevent damage.

IAU's Concern

The International Astronomical Union echoed the widespread concern in the astronomical world and, on June 3, 2019, stated:

> The International Astronomical Union (IAU) is concerned about these satellite constellations. The organization, in general, embraces the principle of a dark and radio-quiet sky as not only essential to advancing our understanding of the Universe of which we are a part, but also as a resource for all humanity and for the protection of nocturnal wildlife. We do not yet understand the impact of thousands of these visible satellites scattered across the night sky and, despite their good intentions, these satellite constellations may threaten both.

Visiting the IAU web site dedicated to satellite constellations https://www.iau.org/public/themes/satellite-constellations/ the concerns by the astronomical community are spelled out, as follows

> The scientific concerns are twofold. Firstly, the surfaces of these satellites are often made of highly reflective metal, and reflections from the Sun in the hours after sunset and before sunrise make them appear as slow-moving dots in the night sky. Although most of these reflections may be so faint that they are hard to pick out with the naked eye, they can be detrimental to the sensitive capabilities of large ground-based astronomical telescopes, including the extreme wide-angle survey telescopes currently under construction. Secondly, despite notable efforts to avoid interfering with radio astronomy frequencies, aggregate radio signals emitted from the satellite constellations can still threaten astronomical observations at radio wavelengths. Recent advances in radio astronomy, such as producing the

first image of a black hole or understanding more about the formation of planetary systems, were only possible through concerted efforts to safeguard the radio sky from interference.

The IAU is a science and technology organisation, stimulating and safeguarding advances in those areas. Although significant effort has been put into mitigating the problems with the different satellite constellations, we strongly recommend that all stakeholders in this new and largely unregulated frontier of space utilisation work collaboratively to their mutual advantage. Satellite constellations can pose a significant or even debilitating threat to important existing and future astronomical infrastructures, and we urge their designers and deployers, as well as policy-makers, to work with the astronomical community in a concerted effort to analyse and understand the impact of satellite constellations. We also urge the appropriate agencies to devise a regulatory framework to mitigate or eliminate the detrimental impacts on scientific exploration as soon as is practical.

To further promote and safeguard astronomical research, on june 2021 the IAU decided to establish an IAU Centre for the Protection of the Dark Sky from Satellite Constellation Interference. The centre will foster the development of tools and procedures to mitigate the impact of satellite constellations on optical astronomy, and work with the space companies and industries to discuss and converge on mitigations.

11

Who Gave Permission?

Faced with the deterioration of the quality of the sky caused by too many satellites, one wonders how the Federal Communications Commission (FCC) gave SpaceX the green light to launch a constellation of 12,000 new satellites without considering their environmental impact. Indeed, the National Environmental Policy Act (NEPA) prescribes that all federal agencies examine the environmental impact of everything they authorize, but the FCC does have an exemption.

With a current population of 9000 satellites (counting the 1500 that are active and the 7500 that have stopped working and are, for all intents and purposes, wreckage), the FCC did not hesitate to more than double the population of objects in orbit by authorizing the launch of 12,000 satellites. And those are just the ones belonging to SpaceX! We have seen that OneWeb, which originally planned a constellation of 650 satellites, is considering launching 6300, to which we have to add the 3200 of

© The Author(s), under exclusive license to Springer Nature Switzerland AG 2021
P. Caraveo, *Saving the Starry Night*,
https://doi.org/10.1007/978-3-030-85064-7_11

Amazon's Kuiper project, Samsung's 4700, and Boeing's almost 3000. Not to mention the Russian and Chinese plans and the fact that Elon Musk has always said that he wants to expand Starlink to up to 42,000 satellites and has already asked, and obtained, authorization from the FCC to launch another 30,000 satellites.

Let's Talk About Environmental Impact

It is now easy to envisage a future with very busy orbits, which, among other things, amplifies the possibility of collisions between satellites, with chain effects that are worrying, to say the least. In fact, in the event of a collision, a cloud of debris is formed that will continue to travel along the same orbit of the original satellite, multiplying the probability of further impacts. This is called Kessler's syndrome, after the scientist who first studied it. Despite the apparent vastness of circumterrestrial space, we already know that the probability of collision is not zero. On February 10, 2009, one of the satellites of the Iridium constellation collided with a Russian military satellite of the Kosmos series that was no longer operational. In addition to destroying the two satellites, the collision generated a cloud of debris that took months to decay. The population of objects orbiting the Earth (consisting of active satellites, spent satellites and debris of various sizes) is kept under constant surveillance to minimize the risk of dangerous approaches. Of particular concern are the switched-off (or faulty) satellites and debris of various kinds that can no longer respond to commands but continue to travel in their orbit at about 8 km/s. The International Space Station occasionally has to perform manoeuvres to dodge objects that might get too close.

Musk claims that his countless satellites are equipped with an automatic anti-collision system, and no one doubts his word. However, on September 2, 2019, the European Space Agency had to move its Aeolus satellite to avoid a too-close pass with a Starlink. It may have merely been a communication problem between the ESA and SpaceX, however, it was not a good start.

Musk says that his satellites are able to deorbit at the end of their service, so that they destroy themselves upon contact with the atmosphere. This is certainly true, since the satellites have an ion engine for the purpose of varying their orbit; however, some of the Starlink satellites have already stopped operating and are no longer able to control their orbit. After all, they're low-cost satellites, and it's reasonable to expect a failure rate of around 5% that turns them into wreckage in orbit until the very slight, but continuous, friction with the very little residual atmosphere slows down their speed, taking them low enough to be destroyed. In any case, even if everything works fine, there will be continual controlled re-entries, with satellites lowering their orbit and burning in the atmosphere. However, it would be wrong to think that they will disappear without a trace. The superheated material is transformed into toxic gas and dust that will remain suspended in our atmosphere, perhaps diffusing solar radiation and contributing to climate change. The same can be said for emissions from the fuel used for launches. These are all sources of potentially polluting gases in the atmosphere. To get an idea of the dimensions of the problem, let's consider that we are talking about having 100,000 satellites in orbit with an orbital life that can span from 5 to 10 years. This means that, every year, approximately 10,000 satellites will have to be launched to replace those that have reached the end of their life and need to be de-orbited. Assuming that each launch brings 50 satellites into orbit, at least 200 launches per year will be needed just to maintain the internet network. Other launches will then be needed to meet all the other civil and military demands for services, such as Earth observations, weather forecasts, and border surveillance, just to name a few. We're talking about hundreds of other launches that will burn large quantities of fuel, spewing exhaust gas the composition of which depends on the fuel used, but that can contain nitrogen and chlorine oxides that can react with ozone, destroying it and thinning the layer of this gas so important to protecting us from the Sun's ultraviolet radiation. SpaceX uses a kerosene-like fuel that deposits large amounts of blackish dust in the stratosphere that is very efficient at absorbing solar radiation. The warming of the stratosphere accelerates chemical reactions that destroy ozone. Unfortunately, it is possible that some of these effects are already present in the northern hemisphere, from where most launchers depart. Earth's atmospheric experts couldn't explain why, during the winter and spring of 2020, the largest and longest-lasting ozone hole ever recorded developed over the Artic. Could it be due to SpaceX launches? If so, it's time to start worrying.

This too is a potentially significant environmental impact that seems to have gone completely unnoticed, for the very good reason that the FCC is exempt from enforcing NEPA because its decisions *"have no significant effect on the quality of the human environment."*

If it were possible to prove that this is not true, because satellites have an appreciable effect on air pollution and on the quality of the sky, or because they compromise access to space owing to the increased probability of collision, it would be possible to take the FCC to court and ask for a suspension of the launch authorization (in fact, a stop to all planned launches) until a serious assessment of the environmental damage has been made. To this end, the Italian astronomer Stefano Gallozzi has launched a petition, which has been signed by over 2000 professionals in Italy and abroad. The issue is now in the hands of an international team of law firms that is preparing the documentation.

However, astronomers are not the most troublesome critics of SpaceX. I think the most worrisome actions, from SpaceX's perspective, come from other satellite service operators, who fear that their access to space will be severely limited. On December 22, 2020, Viasat officially asked the FCC to do an environmental impact assessment of SpaceX, since, "given the sheer quantity of satellites at issue here, as well as the unprecedented nature of SpaceX's treatment of them as effectively expendable, the potential environmental harms associated with SpaceX's proposed modification are significant."

Viasat knows very well that the FCC has a blanket exemption, but goes on to say:

"Relying on the Commission's decades-old categorical exemption to avoid even *inquiring* into the environmental consequences of SpaceX's modification proposal would not only violate NEPA, but also would needlessly jeopardize the environmental, aesthetic, health, safety, and economic interests that it seeks to protect, and harm the public interest." In mentioning the aesthetic harm, Viasat resonates with the astronomers' request, but it goes on to add that "The Commission's decision thus will directly affect the amount of light pollution in the environment, placing NEPA responsibilities squarely on the Commission's shoulders."

Who knows if the satellite operators, in direct competition with SpaceX, will be able to get the environmental assessment?

For sure, the occupation of low altitude orbital space is growing at an alarming rate. According to the authors of a very recent study focussed on the evaluation of the impact of space activities in low earth orbit, "we have already "filled up" one-third to one-half of the "capacity" of LEO able to sustain long-term space activities as we conceive them now. And it took us 63 years to get to this point. It should therefore be of great concern that in the next decade alone is foreseen the launch of a number of satellites that could equal, at best, but also exceed tenfold, at worst, those put into orbit since the beginning of the space age."

The International Legislative Framework

In parallel, there are also those who raise deeper issues of international law. While it is certainly true that the Starlink satellites are launched from American territory, and so it is reasonable that the authorization is given by an American Authority, it is equally true that their orbits fly over all the other countries of the globe, which have not been involved at all in the decision. On what legal basis does the FCC decide what should go over my head, changing the appearance of my sky in the process?

Therefore, logically, the problem should be managed by the United Nations or, better, by the UNOOSA (United Nation Office for Outer Space Affairs), the world organization, an emanation of the United Nations, that deals with the peaceful use of space.

There is an international treaty on the use of space, the Outer Space Treaty, which says, for example, that no government can proclaim its sovereignty over a celestial body or any part of it, but it says nothing about the occupation of circumterrestrial orbits, the reflectivity of satellites or the safeguarding of astronomy. This is not surprising: the treaty was negotiated in the 1960s and, at the time, the problem simply did not exist. In fact, until a few years ago, going into space was the sole prerogative of space agencies, which launched a limited number of satellites. Now, the situation has changed. Private industries have become protagonists of the new Space Economy, the profits from which are based on the cut of the launch cost and on the multiplication of the number of satellites, with the result that it is much more difficult to keep the situation under control. In order to avoid the dangers from an indiscriminate privatization of space, the Outer Space Treaty should be updated. Indications on the maximum reflectivity of what goes into orbit should be given, so as not to repeat the problem of the light trains, but it would also be advisable to control the number of satellites that can be put into orbit. The fact that a single operator is able to launch thousands and thousands of satellites doesn't mean that this is necessarily the right thing to do. Just as it was decided that the worldwide use of frequencies, treated as a common good of humanity, should be controlled, so it would be necessary to rationalize the use of circumterrestrial orbits to avoid overcrowding with increased probability of collision and the consequent Kessler syndrome, which would make circumterrestrial space unusable, with enormous damage to our satellite-dependent society. In fact, the IAU approached UNOOSA and, together, they organized a congress in October 2020, entitled Dark and Quiet Skies for Science and Society (https://unoosa.org/oosa/events/data/2020/dark_and_quiet_skies_for_science_and_society.html).

This congress was devoted to the discussion of all the issues related to the preservation of the sky as a scientific and cultural heritage and as a very important part of the natural environment where all forms of life that inhabit our planet have evolved and developed.

From the proceedings of the congress, available online, a series of recommendations to be submitted for consideration by the UN Assembly have been prepared. Meanwhile, in October 2021, the astronomical community met again during a highly attended online Conference with the same title to continue discussing the issues realted to the sky safeguard and to learn how to better coordinate the mitigation efforts (https://www.unoosa.org/oosa/en/ourwork/psa/schedule/2021/2021_dark_skies.html).

Negotiating resolutions acceptable to all UN Member States will take time, but this is the only way forward, since actions need to be taken at the global level and the necessary uniform directives formed.

12

Conclusion

The dark sky is a part of humankind's heritage, as well as that of other life forms on this planet.

This doesn't mean that astronomers don't understand the social utility of night-lighting or connectivity for those living in regions that are not covered; they only ask that their efforts, financed with public money, are not thwarted in the name of profit. This is all a consequence of the opening up of space to private companies, an opening that makes the Space Economy flourish, but has a series of side effects that should be regulated.

Elon Musk assures us that he absolutely doesn't want to damage astronomical research, but then he adds that, if astronomical observatories have too many problems to operate from the ground, it would be advisable to place them in orbit or, even better, relocate them to the Moon. Let's just say that this is not an original idea: for years, radio astronomers have been dreaming of being able to

© The Author(s), under exclusive license to Springer Nature Switzerland AG 2021
P. Caraveo, *Saving the Starry Night*,
https://doi.org/10.1007/978-3-030-85064-7_12

build an instrument on the hidden face of the Moon, where it would be possible to operate far from all terrestrial interference. Unfortunately, it would be a difficult and expensive project and, in order not to thwart the effort, it would be immediately necessary to start protecting at least part of the Moon's hidden hemisphere to avoid it being polluted (both from the environmental and radio emission points of view) by the commercial exploitation that seems to be around the corner. Even the idea of using orbiting telescopes doesn't solve all of the problems. The Hubble Space Telescope, for example, orbits at an altitude of about 550 km, and space debris' as well as satellites' trails are typically seen in 5% of its images. it has already happened that its field of view was crossed by a Starlink satellite which left a bright fat strike, as shown in the figure (Fig. 12.1).

And it's bound to happen again, since OneWeb chose orbits at 1200 km for its 6300 satellites and Samsung

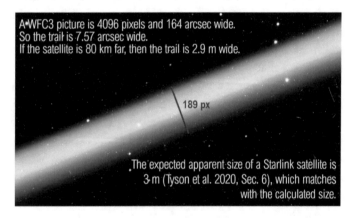

Fig. 12.1 In November 2020 Starlink 1619 crossed the HST field of view while orbiting at about 80 km from the telescope. Thus, its strike appears to be out of focus (*Credit* Jonathan McDowell on Twitter Image from Mikulski archive for Space Telescope)

Fig. 12.2 Interested in the aurora, the astronauts shoot a short video (https://youtu.be/zdL2nE5nwlc), but the Starlink satellites enter the frame (as we see in this still) (*Credit* NASA)

plans a constellation of 4700 satellites placed at an altitude of 2000 km.

For sure, even photos taken from the International Space Station do witness Starlink *photo bombing*, as shown by this frame extracted from a of a beautiful aurora short video shot by astronauts (Fig. 12.2).

Light trains are everywhere.

Unfortunately, the problem cannot be solved with a magic wand. Just as with the management of light pollution, reasonable solutions need to be developed for the purpose of circumscribing the problems caused by satellite constellations and finding acceptable compromises.

What we must absolutely avoid, however, is the misuse of the sky. There are plans to put in orbit large reflective structures that could become mega-billboards. Please, do not do that!

A Note of Thanks

Flaubert said that gratitude does not need to be expressed, however, I think it is fair to acknowledge that this book would not have existed without the enthusiasm of the students who participated in the competition organized by the Agorà association in San Severo di Puglia in February 2020.

In the end, one of the teams that had grappled with a problem related to the impact of mega-constellations on astronomical research won the competition, thanks to the vote of the large audience present. The credit was all theirs, but it had shown me how interesting and engaging the topic could be for the general public.

My friend Claudia Coga liked the idea very much and decided that my book should inaugurate the new series "Le grandi voci" of the publishing house Dedalo.

However, writing the book seemed like a mission impossible. The timetable I had been given didn't fit in with my busy travel schedule. Then came the pandemic, and the confinement at home forced me to find the time

© The Editor(s) (if applicable) and The Author(s), under exclusive license to Springer Nature Switzerland AG 2021
P. Caraveo, *Saving the Starry Night*,
https://doi.org/10.1007/978-3-030-85064-7

to write my declaration of love to the starry night. The book, entitled **Il cielo è di tutti,** was published in July 2020.

The information on the Italian regulations comes from Fabio Falchi, who, together with Pierantonio Cinzano, has been fighting for years to raise awareness in local administrations about promoting a choice of lighting that respects the darkness of the night. I am grateful to Fabio, who shared his meritorious work and provided me with several pictures.

I would also like to thank Stefano Gallozzi, who kept me continually informed about the initiative he is coordinating. The data on the study conducted during the lockdown in Veneto were provided to me by Sergio Ortolani, while Federico di Vruno updated me on the problems faced by radio astronomers.

A generalized thank you goes to all the friends who, while forced to stay home, read the text in advance, helping me to improve it.

A special thank to my daughter Giulia, who contributed with her memories to the description of satellite hunting, an activity that she loved to do with her dad from the terrace of our house by the sea in the summer evenings.

The preservation of the sky as a priceless cultural heritage interested another friend, Marina Forlizzi, who convinced me to expand the text in view of the English version. I took advantage of what I learned during the online symposium organized in October 2020 by the IAU and UNOOSA and then, again, on Oct 2021. Discovering the biological effects of ill-advised and excessive artificial light took me by surprise, and I decided also to cover this subject, which was new to me.

Again, the writing was done during a lockdown, that of early 2021. In fact, the decrease in mobility imposed by the pandemic has caused us to understand very well the importance of individual behaviour vis-à-vis atmospheric and light pollution, as well as seismic noise produced by a humanity continually on the move. These are important lessons that I hope we will remember when we will be able to move freely again, perhaps to reach astronomical observatories on the top of a mountain in the Andes or simply to go for a walk in search of a dimly lit place to enjoy the most beautiful vision nature can offer us: the starry sky.

Printed in the United States
by Baker & Taylor Publisher Services